Y0-AWH-792

FUEL FIELD MANUAL

FUEL FIELD MANUAL
Sources and Solutions to Performance Problems

Kim B. Peyton

NALCO/EXXON
ENERGY CHEMICALS, L.P.

McGRAW-HILL
New York   San Francisco   Washington, D.C.   Auckland   Bogotá
Caracas   Lisbon   London   Madrid   Mexico City   Milan
Montreal   New Delhi   San Juan   Singapore
Sydney   Tokyo   Toronto

Library of Congress Cataloging-in-Publication Data

Peyton, Kim.
    Fuel field manual : sources and solutions to performance problems /
Kim B. Peyton. Nalco/Exxon Energy Chemicals, L.P.
      p. cm.
    Includes Index.
    ISBN 0-07-046572-X (alk. paper)
    1. Motor fuels.   I. Nalco/Exxon Energy Chemicals, L.P.
Title.
    TP343.P45   1997
    665.5′3—dc21   97-16078
                                                                                  CIP

McGraw-Hill

A Division of The McGraw·Hill Companies

Copyright © 1998 by The McGraw-Hill Companies, Inc. All rights reserved. Printed in the United States of America. Except as permitted under the United States Copyright Act of 1976, no part of this publication may be reproduced or distributed in any form or by any means, or stored in a data base or retrieval system, without the prior written permission of the publisher.

1 2 3 4 5 6 7 8 9 0 DOC/DOC 9 0 2 1 0 9 8 7

ISBN 0-07-046572-X

The sponsoring editor for this book was Robert Esposito, the editing supervisor was Frank Kotowski, Jr., and the production supervisor was Clare Stanley.

It was set in Times Roman by Marion Graphics.

Printed and bound by R. R. Donnelley & Sons Company.

McGraw-Hill books are available at special quantity discounts to use as premiums and sales promotions, or for use in corporate training programs. For more information, please write to the Director of Special Sales, McGraw-Hill, 11 West 19th Street, New York, NY 10011. Or contact your local bookstore.

                This book is printed on recycled, acid-free paper containing a
                minimum of 50% recycled, de-inked fiber.

Information contained in this work has been obtained by The McGraw-Hill Companies, Inc. ("McGraw-Hill") from sources believed to be reliable. However, neither McGraw-Hill nor its authors guarantee the accuracy or completeness of any information published herein and neither McGraw-Hill nor its authors shall be responsible for any errors, omissions, or damages arising out of use of this information. This work is published with the understanding that McGraw-Hill and its authors are supplying information but are not attempting to render engineering or other professional services. If such services are required, the assistance of an appropriate professional should be sought.

Contents

Preface vii

Introduction ix

Acknowledgments xiii

Chapter 1. A Problem Solving Technique 1

Chapter 2. Refining Processes Used in Fuel Production 5

Chapter 3. Critical Properties of Crude Oil and
Common Hydrocarbon Fuels 25

Chapter 4. Common Sources of Fuel Performance Problems 65

Chapter 5. Utilizing Physical and Chemical Property
Measurements to Identify Sources of Fuel Problems 107

Chapter 6. Solving Fuel Problems by Using Chemical Additives 133

Chapter 7. Test Methods Used to Identify and Solve Fuel Problems ... 169

Chapter 8. Identifying and Solving Specific Fuel Problems 193

Chapter 9. Components of Fuel and Fuel Additive Storage
and Injection Systems 211

Chapter 10. Safe Shipping and Hazard Information for
Common Fuels, Oils, and Solvents 233

Chapter 11. Fuel Performance Property and
Problem Solving Guide 245

References 263

Appendices 273

Useful Terms and Definitions 289

Useful Calculations, Conversions, and Formulas 309

Index 311

Preface

Technical support could be defined as the use of practical and theoretical knowledge to solve everyday problems. The problems faced by someone involved in providing technical support are often not clearly defined. Because of this, it is usually necessary to gather as much information as possible related to a problem.

Detailed and accurate analytical analyses are crucial to the success of any technical support or problem solving situation. However, unless it is interwoven within a "common sense" framework, accurate analytical information is of minimal value.

In order to become proficient in providing high quality technical support to the fuel industry, certain components of education and experience are needed. Of primary importance would be:

- An understanding of chemical interactions.
- A mechanical aptitude (how things work).
- An understanding of what information is critical and what is not.
- A knowledge of analytical test methods and their capabilities.
- An ability to visualize how component parts of a system affect the operation of an entire system.
- An ability to ask the "right questions" needed to gain important information.

Gathering and collecting information is important, but alone it will usually not result in a solved problem. All of the preceding factors outlined above must blend to form a scenario describing how a problem or situation developed.

Developing the skill of providing good technical support takes time and experience. The basic components of a broad ranged education and a natural common sense are needed. The chapters in this book will deal with some of the basic facts and physical properties of fuels. The final chapters will focus on common causes of fuel handling and performance problems and how they may be solved.

Kim B. Peyton

Introduction

I wrote the *Fuel Field Manual* as a guide to help others who are faced with solving fuel related problems. The information presented is a blend of both practical and theoretical concepts and is presented in a condensed format for ease of use. I tried to review most of the common problems encountered when fuels such as gasoline, jet fuel, diesel fuel, heating oil, marine fuels and residual fuel oils are refined, transported, stored and used.

It is not necessary to read this manual from cover to cover in order to use it effectively. The various chapters stand alone as special topics for use when needed. However, I hope that the reader will find that all of the information provided will be helpful when working to identify and solve fuel problems.

Individuals who have varying levels of interest and responsibility in developing solutions to fuel problems could use this book. This audience would include petroleum chemists, refining engineers, automotive and truck fleet maintenance personnel, fuel marketers, fuel and fuel additive sales personnel, fuel testing and inspection companies, technical consultants, students and other individuals involved in various aspects of the petroleum industry.

The initial chapters are organized to provide a background of information on problem solving, fuel refining, fuel specifications and fuel properties. Common fuel problems and solutions to specific problems are then discussed. Final chapters contain "hard to find" information about equipment used to store, handle and distribute fuel properly and safely. A chapter-by-chapter summary is presented in the next few paragraphs.

Chapter 1, A Problem Solving Technique, discusses the process developed by Charles H. Kepner and Benjamin B. Tregoe in their book entitled *The New Rational Manager*. An example of how this process can be applied to an actual fuel problem solving situation is presented.

Chapter 2, Refining Processes Used in Fuel Production, presents a brief review of refining processes used in fuel production. The focus of the discussion is on how refinery unit operating parameters can affect finished fuel quality and performance.

Chapter 3, Critical Properties of Crude Oil and Common Hydrocarbon Fuels, provides information on the chemistry of crude oil with a focus on the components which most impact finished fuel quality. A description of the physical properties and specifications of automotive gasoline, aviation gasoline, jet fuel, diesel fuel, fuel oil, marine fuel, burner fuel and residual fuel are reviewed.

Chapter 4, Common Sources of Fuel Performance Problems, details how elements of the environment such as water, air and cold temperature initiate fuel problems. Also, discussed in detail are problems due to fuel wax, olefins, low volatility and microorganisms. A thorough explanation of fuel cetane number and cetane index determinations is also given.

Chapter 5, Utilizing Physical and Chemical Property Measurements to Identify Sources of Fuel Problems, focuses on describing the impact fuel physical and chemical properties can have on the handling and performance of fuels. By understanding the role of these properties in controlling fuel quality and performance, it becomes easier to identify the cause of a fuel problem.

Chapter 6, Solving Fuel Problems by Using Chemical Additives, presents a brief review of common additives used to affect and improve the handling and performance properties of fuels. Also presented is a discussion of limitations and problems which have been identified with the application and use of certain fuel additives.

Chapter 7, Test Methods Used to Identify and Solve Fuel Problems, contains a summary of several of the more helpful test methods used to rate fuel quality and performance. Accompanying the test summary is a brief discussion on how to interpret the results obtained from the testing.

Chapter 8, Identifying and Solving Specific Fuel Problems, describes actual field problems which have occurred and provides possible solutions to each of the problems. Many of the problems presented have more than one potential solution. A choice of both physical methods and chemical methods which can be used to resolve problems is provided.

Chapter 9, Components of Fuel and Fuel Additive Storage and Injection Systems, is a collection of information describing the hardware and equipment used in the storage, handling, and injection of fuel. Often, an understanding of the composition and function of the equipment used in the fuel distribution system can be of significant benefit when analyzing a fuel problem.

Chapter 10, Safe Shipping and Hazard Information for Common Fuels, Oils and Solvents, describes some of the international guidelines and regulations pertaining to shipment of hazardous materials such as fuel. Also, regulatory information for selected fuels and fuel components is provided.

Chapter 11, Fuel Performance Property and Problem Solving Guide, is a quick-reference guide which summarizes much of the information described in the earlier chapters. It contains a listing of common fuel problems, possible causes for each problem and potential solutions. Steps to take in order to prevent these problems are provided.

Included in the final pages are references, useful terms, definitions, and appendices. This information can be used to help clarify discussions and details presented in this book.

The index is organized so that key words related to fuel problems are pinpointed. Examples include 11 entries to "gum formation," 15 entries to "emulsions" and 15 entries to "rust." Index references are probably the most effective way to use this book to find information needed to help identify and solve problems.

I believe you will find the *Fuel Field Manual* to be unique because the information presented covers a wide range of topics related to fuel properties, storage, handling and use. This broad scope focus enables the reader to look beyond fuel properties alone for the cause of a problem. Although fuel properties can be the source of a problem, often the cause can be linked to an external contamination source or to the impact factors in the physical environment will have on fuel behavior. These external factors can create problems even in perfectly refined and blended fuels and oils.

I owe special appreciation and thanks to ASTM for allowing me to use several test method summaries in the *Fuel Field Manual*. The summaries were extracted with permission from the *Annual Book of ASTM Standards*, copyright American Society for Testing and Materials, 100 Barr Harbor Drive, West Conshohocken, Pennsylvania 19428.

Acknowledgments

First, I want to thank my family for their support and understanding throughout the past four years while I was preparing this manuscript. The many evenings and weekends spent researching, typing and refining this work were made easier by their encouragement.

The continual, positive support provided by the management of Nalco/Exxon gave me the confidence needed to complete and finalize the *Fuel Field Manual*. I also appreciate the many patient hours spent by Susan Bourgain and Joan Wendt working with me, the staff at Marion Graphics and the editors at McGraw-Hill to produce this book. Finally, I am grateful for the inspiration of Dr. John Lockhart whose words helped to guide me in this effort.

CHAPTER 1
A Problem Solving Technique

Beginning the process of solving a problem can sometimes be the most difficult part of the process. In many circumstances more information is needed than is available. The painless decision to bypass the tedious process of collecting information is easy to justify. The time required to look closely at all aspects of a problem before recommending a solution is often not available. Problem diagnosis and solutions, however, are often expected readily.

How then are many problems solved? Often they are solved by individuals with experience in facing and resolving similar situations. Even if all of the information required becomes available, conclusions may be difficult to reach unless an experienced eye fits the pieces of information into a realistic picture.

Obviously, gaining experience in solving problems is not achieved quickly. However, there are techniques which can be used to help focus on the most important components of a problem and to eliminate other factors.

Described below is a technique taken from the book written by Charles H. Kepner and Benjamin B. Tregoe entitled *The New Rational Manager*, Princeton Research Press, 1981. This straightforward process of problem analysis is excellent in providing steps to help identify and focus on the vital information needed to identify and solve problems.

Kepner - Tregoe Problem Analysis Technique

IDENTITY
    WHAT is observed ?   WHAT COULD BE but IS
                                      NOT observed ?

LOCATION
    WHERE IS the problem observed ?   WHERE COULD the
                                      problem be but IS NOT
                                      observed ?

TIMING
    WHEN IS the problem observed ?   WHEN COULD the problem
                                      be but IS NOT observed ?

MAGNITUDE
    HOW MANY occurrences are observed ?   HOW many occurrences
                                          COULD BE but ARE NOT
                                          observed ?

If answers to the above series of questions can be obtained, it is relatively easy to focus in on the actual cause and source of a problem. The strength of the **Kepner - Tregoe** technique is that it helps to eliminate much of the "hearsay" information and will fortify the relevant information.

The following example describing a gasoline contamination problem may be used to clarify the effectiveness of the **Kepner - Tregoe** technique:

Problem Scenario

A retail marketer of gasoline received numerous complaints from customers regarding engine shutdown after fueling at the retailer's service station. Complaints came from customers who fueled with premium grade gasoline. Complaints were all received within a two week time frame in January. Below freezing overnight and daytime temperatures were common throughout this same two week time span in January.

The fuel retailer could not pinpoint the source of the problem since complaints were coming from customers who purchased premium grade fuel at different stations throughout the area.

The problem began shortly after the retailer switched to a new gasoline additive for use in both regular and premium grade gasoline. Immediately, the gasoline additive supplier was contacted by the retailer to help resolve this problem. In the retailer's eyes, the customer complaints began immediately after initiating use of the new fuel additive. Therefore, the additive was probably the cause of the engine shutdown problem.

To resolve the problem, personnel from the gasoline additive supplier worked with the customer to collect service station gasoline samples, customer fuel filters, and samples of gasoline from the retailer's local service stations and from the fuel distribution terminal. All of these samples were thoroughly analyzed. Also, January fuel delivery records were closely audited.

From the information collected, the following observations were made:

WHAT is observed ?
- Engine shutdown was observed in vehicles using premium unleaded gasoline.
- All gasoline was distributed to local service stations from a single fuel terminal.
- Water was found on the bottom of several service station fuel tanks. Water contained high levels of iron (rust), magnesium, sodium and calcium.
- Mg:Ca ratios were close to 3:1. These Mg:Ca ratios are typical of sea water, not fresh water.

- Iron (rust), magnesium, silica, sodium and calcium were identified on customer fuel filters.

What COULD BE but IS NOT observed ?
- Problems in vehicles using regular unleaded gasoline were not observed.
- Gasoline additive was not observed on customer fuel filters.

WHERE IS the problem observed ?
- Interviews with customers revealed that fuel purchases were from 15 service station locations.

WHERE COULD the problem be but IS NOT observed ?
- Problems were not reported from customers who fueled from 140 of the retailer's remaining service stations.

WHEN IS the problem observed ?
- Within a two week time period in January.

WHEN COULD the problem be but IS NOT observed ?
- Continuously

HOW MANY occurrences ?
33

HOW MANY occurrences COULD BE but ARE NOT observed ?
>5000

OBSERVATIONS FROM DISTRIBUTION TERMINAL TANK RECORDS

An oceangoing barge delivery of premium unleaded gasoline was off-loaded into terminal storage 4 days before the first customer complaint was received. Storage tank records indicate that the 1250 gallons of water settled from the fuel after 24 hours in storage. Present tank records show a water level of zero gallons in the terminal storage tank.

Premium unleaded gasoline was delivered from this terminal to over 160 different service station locations during the two week time period that customer complaints were received. Two additional barge loads of fuel were off-loaded into the same terminal storage tank during this two week period.

From the information collected, the following conclusions can be made:

- Engine shutdown occurred within a two week time period in January.
- Problem fuel was purchased from a limited number of stations.
- All fuel was distributed from a single terminal source.

- Fuel off-loaded from the barge contained significant levels of water.
- Significant levels of water are no longer observed in the terminal tank.
- Service station gasoline tanks contain water with a chemical composition (Mg and Ca) similar to sea water. Iron (rust) and sodium were also found.
- Metal contaminents found on customer fuel filters are similar to those found in service station tank water.

CAUSE OF THE PROBLEM:

Water (Ice?) and Corrosion Products in Fuel Lines

Sea water and corrosion products were delivered with premium gasoline by barge to the fuel distribution terminal storage tanks. The water settled to the bottom of the terminal tank and was drawn off with fuel and delivered to service station tanks. Customers fueling at these service stations pumped premium gasoline containing sea water and corrosion products into their automobiles.

The fuel filters were then partially blocked by rust and corrosion products carried by the water. Below freezing temperatures possibly initiated the formation of ice within fuel filters and in fuel lines containing water. The combination of ice and corrosion products blocked fuel flow through vehicle fuel lines and filters. Gasoline flow to the engine was halted and engine shutdown followed.

Not all service stations receiving premium unleaded fuel from the distribution terminal were affected. Only a limited number of stations received fuel containing water and corrosion products were affected. This explained the random nature and limited scope of the vehicle problems.

By using the **Kepner - Tregoe** approach to problem solving, information can be collected in a manner which seems to "focus in" on only the critical details of a problem. Once collected, the information can be compiled into a format which enables one to develop logical conclusions and to determine the cause of a problem.

CHAPTER 2

Refining Processes Used in Fuel Production

The processes utilized in the refinery to produce finished fuels such as gasoline, jet fuel, diesel fuel and heating oil can be quite complex. Constant monitoring of the refining process is required by engineers. This is because the characteristics of the various crude oils refined can change frequently and also because the finished product volume demand can change often.

To fulfill finished product blend requirements, it is imperative that all processes within the refinery function and work together as a continuous system. An upset or problem with the operation of one unit in the refinery can sometimes dramatically affect the efficiency, output and quality of the entire refining process.

A brief description of the various units found within the typical fuel refining operation is provided in this chapter.

A. DISTILLATION

Fuels and other petroleum products are derived from crude oil through the use of a variety of different refining process techniques. Distillation, however, is the first significant processing step taken in crude oil refining. Both atmospheric and vacuum distillation can be utilized to process crude oil into fuels and other products.

Desalting

Before distillation, crude oil salts and certain metals must be removed. The process of desalting is applied for this purpose. Desalting involves mixing the crude oil with water at a temperature of about 250°F (121.1°C) under enough pressure to prevent evaporation of both water and volatile crude oil components. The salts are dissolved and removed by the water. Oil/water emulsions often form which also contain salts. The emulsions can be broken by the use of high voltage electrostatic coalescers or by the use of demulsifying chemicals.

Single and multi-stage desalting processes are utilized to remove as much of the salt as possible. Any remaining salts which carry into the distillation unit could initiate corrosion.

From the desalting unit, crude oil passes through a series of heat exchangers to increase the temperature to about 550°F (287.8°C). From the exchangers, oil passes into a furnace or pipe-still and heated to about 650°F to 700°F (343.3°C to 371.1°C). The heated oil then enters the atmospheric column.

Once in the column, volatile compounds vaporize. Reflux begins and oil fractions condense and collect in various trays throughout the column. Heating within the column can influence the rate of reflux and holdup within the column. Steam is introduced into the column at the level of the lower trays and strips lighter compounds from the condensed fractions on these trays. As a result, the flash point of the fractions remaining on the lower trays increases.

1. Atmospheric Distillation

Most atmospheric columns contain from 30 to 50 fractionation trays. For each sidestream desired, about five to eight trays are required, plus additional trays above and below the primary trays. The sidestreams collected from the distillation column usually contain light end gases which must be stripped and removed from the collected fractions. Smaller stripping towers are used to remove these light-end products. Steam-stripping is typically used for this process.

Light straight run (LSR) gasoline containing pentanes and other compounds with boiling points less than about 180°F (82.2°C) are condensed at the overhead. Some of this LSR gasoline returns to the column as reflux, but most is separated and collected. It is then further processed through desulfurization or isomerization.

The naphtha or heavy straight run (HSR) gasoline fraction is usually collected from the middle trays of the atmospheric column. This fraction can be used as gasoline blendstock or in reformate production.

A gas oil fraction used in the production of kerosene and distillate fuel is collected in the lower trays of the atmospheric column. The atmospheric tower bottoms are sent to the vacuum tower for distillation or can be further processed by cracking into lighter fractions.

2. Vacuum Distillation

Removal of naphtha and distillate fractions from the crude oil under atmospheric pressure distillation requires charge temperatures to be maintained below the cracking temperature of the crude oil components. This temperature will vary but can typically range from 700°F to 750°F (371.1°C to 398.9°C). Occasionally, even lower temperatures may be required. Above these temperatures, crude oil components can begin to thermally crack and foul processing equipment.

Vacuum distillation utilizes reduced pressures and temperatures as a means to remove remaining fuel components and higher boiling lubricant fractions from atmospheric tower bottoms. In the vacuum unit, atmospheric distillation tower residual components are further distilled without being subjected to the high temperature conditions which lead to cracking and degradation of oil components. Under absolute pressure levels ranging from 25 to 120 mmHg, vacuum distillation will remove higher boiling materials effectively.

In order to minimize the tendency of coke formation within the column, steam stripping is utilized. Steam utilization also helps to reduce the absolute pressure of the system to 10 mmHg or less and can help stabilize the desired unit vacuum levels. By operating at low vacuum pressures, the product yield will increase and the operating costs will typically be reduced.

The vacuum distillation unit charge heating temperatures are dependent upon the operating vacuum pressure and the IBP of the charge. However, the unit operating temperatures are always maintained low enough to prevent thermal cracking of the charge. Upon removal of useful fuel and lubricant components from the vacuum unit charge, high molecular weight, high boiling compounds remain as a vacuum residual fraction. The vacuum residual can vary in composition from refinery to refinery and sometimes from day to day within the same refinery if a variety of crude oil blends are processed.

Typical products obtained from the atmospheric and vacuum distillation of crude are briefly described in **TABLE 2-1**.

TABLE 2-1. Typical Products Obtained from Atmospheric and Vacuum Distillation of Crude Oil

Product	Description
Fuel Gas	Primarily methane and ethane; sometimes propane; often referred to as "dry gas"
Wet Gas	Primarily propane and butane; can also include methane and ethane; propane and butane are used in LPG; butane as gasoline blendstock
Light Straight Run Gasoline	After desulfurization, this stream is used in gasoline blending or as isomerization feedstock
Heavy Straight Run Gasoline or Naphtha	Utilized primarily as catalytic reformer feedstock
Gas Oils	Utilized as straight run distillate after desulfurization. Lighter atmospheric and vacuum gas oils are often hydrocracked or catalytically cracked to produce gasoline, jet and diesel fuel fractions; heavy vacuum gas oils can be used to produce lubestocks
Residual Oil	Vacuum bottoms can be utilized as feedstock for the visbreaker, coker or possibly the asphalt unit

B. THERMAL CRACKING

By using elevated temperatures, high boiling components of crude oil distillation can be converted into lighter fractions for use as gasoline and distillate fuel blendstock. Heating residual oil to temperatures between about 800°F and 1000°F (426.7°C and 537.8°C) at increased pressures, carbon-carbon bonds break to yield free radicals. This process of breaking carbon-carbon bonds is called "cracking."

$$C_{14}H_{26} \rightarrow 2CH_3(CH_2)_5\overset{\bullet}{C}H_2$$

Cracking of 14 Carbon Atom Paraffin to Form Two 7 Carbon Radicals

At pressures <100 psi and temperatures >1000°F (>537.8°C), cracking reactions take place in the vapor phase. The formation of lower molecular weight, gaseous hydrocarbons is favored under these conditions.

At pressures between 250 psi to 1000 psi, and temperatures between approximately 750°F and 900°F (398.9°C and 482.2°C), cracking reactions take place in the liquid phase. Gasoline and distillate type products are formed under these cracking conditions.

Hydrogen transfer, addition and dehydrogenation reactions occur among the produced radical species to yield new compounds. These reactions are summarized in **TABLE 2-2**.

TABLE 2-2. Effect of Process Variables on Thermal Cracking Unit Products

Reaction	Description
Hydrogen Transfer	Radical reacts with a hydrocarbon; H-transfers from hydrocarbon to radical forming a lower molecular weight compound and a new radical
Addition	Free radical reacts with olefin to form higher carbon number, sometimes branched, free radical; radical may further crack or undergo H-transfer
Dehydrogenation	Dehydrogenation to form a diene can occur under severe thermal cracking conditions

Naphthenes may undergo ring cleavage or side chain removal when thermally cracked. Longer side chain naphthenes are the most susceptible to side chain removal by thermal cracking. Aromatic compounds are the most resistant to thermal cracking conditions. The ease by which various hydrocarbons are cracked thermally can be ranked as follows:

EASIEST TO CRACK
              n-paraffins
↓ iso-paraffins
              $cyclo$-paraffins
↓   Aromatics
              Aromatic/naphthenes
              Polynuclear aromatics
DIFFICULT TO CRACK

Due to the tendency of thermal cracking to produce coke as a by-product of the process, most refiners have replaced thermal cracking units with catalytic cracking processes. However, in certain parts of the world, thermal cracking units still exist.

C. VISBREAKING

This process is used to reduce the viscosity of heavy residual fuel oils by heating the oil to temperatures of about 850°F to 950°F (454.4°C to 510.0°C) under pressures between 60 to 300 psi for only a few minutes. This modified version of thermal cracking yields a lower viscosity product by cracking heavier products to lighter components. Approximately 10% to 25% of the residual material is converted to lighter components. Some refiners recover the cracked components for use as gasoline and diesel blending stocks. Frequently though, these lighter components remain in the oil to reduce the viscosity of the residual fuel.

Visbreaking severity is monitored to help minimize cracking and alteration of the nature of asphaltenes within the visbreaker feed. Paraffinic side chain cracking, asphaltene reformation or destruction of the asphaltene-resin complex may occur during visbreaking operations and may result in precipitation of asphaltenes from solution. Asphaltene precipitation has been seen especially when visbroken material is blended with lighter viscosity, paraffinic fuels.

D. FLUID CATALYTIC CRACKING (FCC)

This extremely valuable refining process is utilized extensively to upgrade lower quality petroleum fractions into valuable components, primarily gasoline grade stocks. The fluid catalytic cracking process involves preheating a heavy gas oil such as an atmospheric gas oil (AGO), vacuum gas oil (VGO) or other residual products to temperatures between about 600°F to 850°F (315.5°C to 454.4°C). The feed is then introduced into the reactor accompanied by steam. Average reactor temperatures can range from 850°F to 950°F (454.4°C to 510.0°C). Temperatures are varied to control octane number and component distribution.

This feed is mixed with hot catalyst in a reactor line called a riser. Heat from the catalyst and from exothermic cracking reactions which take place act together to vaporize the oil and allow the mixture to flow up the riser and into the reactor. At this point, most of the cracking reactions have occurred.

The light oil vapors and the catalyst then pass through a structure called the cyclone. This horn-like structure facilitates the separation of the oil vapors from the catalyst. The oil vapors swirl through and out of the top of the cyclone and into the fractionator. Most of the catalyst particles do swirl out of the cyclone, but collect onto the cyclone walls and drop down the bottom of the FCC reactor.

1. FCC Product Fractions

In the fractionating tower, the cracked oil is distilled into the following components:

Propane-Butane - These gaseous fractions are collected along with other compounds such as hydrogen sulfide and mercaptans. Further processing by washing with compounds such as monoethanolamine or caustic removes hydrogen sulfide and mercaptans. Propene can be used as chemical feedstock. Butene and isobutane can be directly utilized in gasoline blending.

FCC Gasoline - This fraction typically contains a mixture of paraffins, olefins and aromatic compounds in a general ratio of around 5:3:2. This ratio will often vary depending upon feedstock, catalyst quality and reactor parameters. The research octane number of FCC gasoline can typically have values much higher than the motor octane number.

Light Cycle Oil - Heavier compounds formed during FCC processing can reflux and accumulate within the FCC unit. This accumulated fraction is called *light cycle oil* (LCO) and contains high percentages of monoaromatic and diaromatic compounds plus olefins and heavier branched paraffins. LCO is quite unstable and has a very low cetane number. For this reason, it is blended into distillate fuel at relatively low amounts.

FCC Residual / Heavy Cycle Oil / Slurry Oil / Decant Oil /Clarified Oil - This heavy product is primarily aromatic in nature and frequently has a density >1.0. It can serve as a heavy fuel oil blendstock or possibly as a carbon black oil feedstock after catalyst fine removal.

During the cracking process, carbon deposits or coke build up on the spent catalyst particles. These deposits can deactivate the catalyst performance and must be removed. This is typically accomplished in two stages. First, the catalyst collected at the bottom of the reactor is steam stripped to remove residual hydrocarbon. The stripped catalyst then passes into the regenerator and is heated with air to temperatures as high as 1,100°F to 1,200°F (539.3°C to 648.9°C). At these temperatures, coke burns off of the catalyst making it ready for reuse within the FCC unit.

Percent conversion in an FCC unit is based upon the amount of material collected which boils below a temperature of 430°F (221.1°C). Components with boiling points above 430°F (221.1°C) reduce the yield of the FCC unit. For example, if 20% of the cracked components have boiling points >430°F (>221.1°C), the percent conversion is calculated to be 80% (100 - 20 = 80).

The effect of catalytic cracking on various hydrocarbon feedstock components is described in **TABLE 2-3**.

TABLE 2-3. Products Produced from Catalytic Cracking of Different Hydrocarbon Types

Feedstock Component	Effect of Catalytic Cracking
Paraffin	High concentrations of C_3 and C_4 hydrocarbons are produced; olefins and aromatics are also produced; molecules containing greater than six carbon atoms are cracked more readily than molecules with fewer carbon atoms
Olefin	Crack at higher rates than paraffins; olefins isomerize and aromatize to yield higher octane components
Naphthenes	Side chains are broken free of ring structure; dehydrogenation of naphthenic rings containing nine or more carbon atoms is common
Aromatics	Rings possessing three or fewer carbon atoms attached to the ring are quite resistant to cracking; long hydrocarbon chains are usually broken and freed from the aromatic ring

2. FCC Cracking Reactions

Catalytic cracking proceeds through the formation of carbocations. The ease of formation depends upon the structural nature and stability of the carbocation generated.

Easiest to Form
↓   Tertiary
           Secondary
↓   Primary
           Ethyl
↓   Methyl
Difficult to Form

Once formed, the carbocations undergo a variety of reactions including β-scission, isomerization, hydrogen transfer and termination.

a. β-scission

In this reaction, carbocation cracking occurs at the C-C bond which is β in position relative to the carbon possessing the positive charge. This type of cracking yields relatively large amounts of C_3, C_4, C_5 and higher carbon number olefins. The carbocation which remains continues to crack, isomerize and undergo further β-scission. The reaction is usually terminated at some point by hydrogen transfer from another molecule.

b. Isomerization

Hydride or methyl group shift to form the more stable carbocation species occurs during catalytic cracking. For example, methyl group shift to form a tertiary carbocation from a secondary ion species is favored. The presence of relatively high concentrations of branched species in FCC products can be explained by this mechanism.

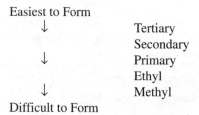

        Secondary Carbocation   Tertiary Carbocation

E. HYDROTREATING OR HYDROPROCESSING

This refinery process is used extensively to improve the quality and usefulness of petroleum products. The primary use of hydrotreating is for the reduction of olefins to paraffins and the removal of sulfur from products. Hydrotreating also effectively removes nitrogen and oxygen heteroatoms and can reduce aromatics to naphthenes and/or paraffins.

The process involves mixing oil with hydrogen under pressure and passing this mixture through a heated catalyst bed. Catalysts such as cobalt-molybdenum (CoMo) oxides on alumina or nickel-molybdenum (NiMo) compounds on alumina are commonly used. The CoMo catalysts are effective at reducing sulfur and are commonly used throughout the refining industry to produce low sulfur products. NiMo catalysts are effective at reducing both fuel sulfur and nitrogen compounds, but often require careful application in the unit.

Effective hydrotreating removes sulfur, nitrogen and oxygen from products yielding H_2S, NH_3 and H_2O, respectively. Incomplete hydrotreating can produce mercaptans, amines and alcohols. Although rare, carryover of these compounds into finished products can lead to copper corrosion, stability or odor problems.

Sulfur removal through hydrotreatment is used in the following process applications:

- Production of low sulfur kerosene and low sulfur diesel fuel.
- Reduction of residual fuel oil viscosity and sulfur content to meet specifications.
- Reformer naphtha charge sulfur reduction.

Examples of desulfurization through hydrotreatment are shown in **FIGURE 2-1.**

Figure 2-1. Desulfurization of Fuel Organic Compounds Through Hydrotreatment

$$RSH + H_2 \rightarrow RH + H_2S$$

benzothiophene + $3H_2$ ⟶ ethylbenzene + H_2S

F. CATALYTIC HYDROCRACKING

The hydrocracking unit is typically used to produce high octane gasoline as well as distillate and naphtha fractions. Feedstock consists of cycle oils, coker distillates and other distillate range fractions. These streams are not readily cracked through the fluid catalytic cracking process, but are effectively cracked under the more severe hydrocracking process conditions.

Most hydrocracking processes involve using a single or dual stage, fixed-bed type reactor. The reaction occurs when feedstock and hydrogen are mixed at pressures of 1,000 to 2,000 psi and then are passed into a 500°F to 800°F (260.0°C to 426.7°C) reactor heating zone. If the unit is a dual stage unit, fractions boiling at temperatures <400°F (204.4°C) such as gasoline and naphtha streams are collected from the unit first. The heavier fractions then pass into the second stage reactor and are hydrocracked to yield jet fuel, diesel fuel and other distillates.

The hydrocracking catalysts are typically alumina-silica supporting a rare earth hydrogenation catalyst. Rare earth metals can include nickel, platinum, palladium and tin. Hydrocracking catalysts are poisoned by metals, oxygen and certain organonitrogen and sulfur compounds. For this reason, feedstock is often hydrotreated prior to hydrocracking. Except for metals, this will result in removal of these hydrocracking catalyst poisons.

Also, water removal from hydrocracker feedstock is essential to help ensure that the catalyst does not dissolve and collapse. Filtration through molecular sieve or silica gel will help to remove water. Some hydrocracking processes, however, can tolerate up to 500 ppm of water in the feedstock.

A description of the effect different hydrocracking parameters can have on the overall process is provided in **TABLE 2-4**.

TABLE 2-4. Effect of Process Variables on Hydrocracking Unit Operation

Process Variable	Primary Effect
Temperature	An 18°F (10°C) temperature increase doubles the reaction rate but does not affect conversion as significantly; temperature increase is typically used to compensate for the loss in catalyst activity
Pressure	Increase in the partial pressure of hydrogen increases the rate of conversion
Space Velocity	An increase in space velocity decreases conversion. To compensate for an increase in space velocity, temperature must be increased
Nitrogen Content	Organic nitrogen decreases catalyst activity and decreases the conversion rate
Hydrogen Sulfide	At low levels, hydrogen sulfide can inhibit aromatic ring saturation. This results in higher octane gasoline and low smoke point jet fuel. At high concentrations, cracking catalyst activity is adversely affected

G. ISOMERIZATION

The isomerization process is utilized to convert light paraffins such as butane, pentane and hexane into higher octane isoparaffins. Isoparaffins have higher octane numbers than normal paraffins of the same carbon number. For example, n-pentane has a research octane number of about 61, and isopentane has an octane number of approximately 92.

The process involves first separating mixed butane compounds by distillation to isobutane and n-butane. The n-butane is then mixed with hydrogen, heated and passed through a reactor containing a platinum catalyst or an HCl activated aluminum chloride catalyst. The n-butane is isomerized to isobutane and separated.

Butane isomerization and pentane-hexane isomerization are the two most important isomerization processes. Isobutane is utilized primarily as alkylate feedstock. Isopentanes and isohexanes have become valuable high octane blending components in gasoline.

H. CATALYTIC POLYMERIZATION

In refining, the polymerization process is utilized to produce high octane gasoline components from three and four carbon olefins. It can also be used for the production of certain alcohols and for reacting olefins with aromatic compounds. Cumene and ethyl benzene can be produced through catalytic polymerization.

Products from the catalytic polymerization process are primarily dimers, trimers, tetramers and oligomers rather than true polymers. However the process name is still used throughout the refining industry.

The process involves reacting butenes and mixtures of propenes and butenes with either a phosphoric acid type catalyst (UOP Process) or a nickel complex-alkyl aluminum type catalyst (IFP Dimersol Process) to produce primarily hexene, heptene and octene olefins. Reaction first proceeds through the formation of a carbocation which then combines with an olefin to form a new carbocation species. The acid proton donated to the olefin initially is then released and the new olefin forms.

Hydrotreatment of the newly formed olefin species results in stable, high octane blending components.

I. CATALYTIC REFORMING

The process of catalytic reforming is utilized for the production of high octane number compounds from lower octane number fractions. It is an endothermic process which requires heat and a catalyst in order to maintain a constant reaction rate. Hydrogen is also produced in significant quantities through the reforming process.

Reformer feedstock is usually from a straight run naphtha fraction with a boiling range from about 180°F to 380°F (82.2°C to 193.3°C). Higher boiling compounds are typically not used due to a tendency to crack during the reforming process and form coke on the catalyst. Lighter reformer feedstocks yield an excessive amount of butane and prove uneconomical.

Examples of reforming reactions include the following:

- Dehydrogenation of cycloparaffins to yield aromatics + hydrogen
- Isomerization of paraffins to yield isoparaffins
- Isomerization of naphthenes
- Dehydrocyclization of paraffins to yield naphthenes + hydrogen
- Hydrocracking of higher carbon number paraffins to lower carbon number paraffins
- Dealkylation of alkyl aromatics to yield benzene and a paraffin

Naphtha with a high cycloparaffin content provides the highest yield and is the easiest to process. Naphtha containing high levels of linear and branched paraffins require more severe processing and generally produce a lower yield. The density, boiling point and octane number values of the reformed compounds are higher than those of the reactants.

Feedstock sulfur is converted into H_2S in the reformer. If not removed, it can serve as a poison to the platinum reformer catalyst and diminish the dehydrogenation and dehydrocyclization reactions. Feedstock nitrogen is converted to ammonia in the reformer. If not removed, it can neutralize the acid sites on the catalyst diminishing the ability of the catalyst to promote isomerization, dehydrocyclization and hydrocracking reactions. Hydrotreatment processes and special absorption techniques are utilized to remove sulfur and nitrogen from the reformer feedstock.

Metals such as arsenic, copper and lead are severe reformer catalyst poisons. Only a few parts/billion are needed to poison the platinum reformer catalyst. Water in the feed will promote hydrocracking reactions and lower reformer and produced hydrogen yield.

The reforming reactor is similar to the hydrotreating and hydrocracking reactors in design. The reaction involves preheating the feed to a temperature of approximately 950°F (510.0°C) and passing the feed through one or more columns containing catalyst. A hydrogen pressure of 200 to 500 psig is typically maintained.

Temperature drop throughout the process requires the constant addition of heat. In reforming processes utilizing more than one reactor, the temperature drop in the last reactor is less severe. The final liquid reformate is separated from the produced hydrogen in a separation drum. Butane is then removed from the reformate by fractionation.

Reforming catalysts are typically palladium on a silica-alumina support. The catalysts are deactivated by feedstock contaminants such as organic nitrogen, sulfur, ammonia and H_2S. For this reason, reformer charge may be hydrotreated to remove these components. Also, any trace metal contaminants will be adsorbed onto the hydrotreating catalyst.

J. ALKYLATION

Products from the alkylation unit are utilized primarily as high octane gasoline blendstock. Octane numbers can range from MON 87-95 to RON 94-100. In the fuel refining industry, the term alkylation refers to the reaction of an olefin with an isoparaffin possessing a tertiary carbon atom. The olefins and paraffins are usually low molecular weight molecules containing three or four carbon atoms. Five carbon atom olefin feedstock is rarely used. The isoparaffin:olefin ratio is kept high to minimize olefin polymerization. Isoparaffin:olefin ratios of 5:1 to 15:1 are common.

The alkylation process is highly exothermic and proceeds through the formation of tertiary butyl carbocation species. Alkylate yields are usually quite high, especially at lower reaction temperatures. High octane number compounds such as isoheptane and isooctane are commonly produced.

Refinery alkylation processes utilize either sulfuric acid or hydrofluoric acid as reaction catalysts. The feedstock for both alkylation processes originates primarily from hydrocracking and catalytic cracking operations. Coker gas oils also serve as feedstock in some applications. The differences and similarities between sulfuric acid alkylation and hydrofluoric acid alkylation are shown in **TABLE 2-5**.

TABLE 2-5. Comparison of Sulfuric Acid and Hydrofluoric Acid Alkylation Processes

Sulfuric Acid Alkylation	HF Alkylation
Reactions are performed at temperatures of approximately 50°F (10.0°C) or lower to minimize polymerization and SO_x formation; a costly refrigeration system is utilized to maintain low temperatures	Temperature is limited to <110°F (43.4°C). Reaction is cooled with water rather than by a refrigeration system
Catalyst volume is equal to the charge volume; acid strength is maintained at about 90%	Catalyst volume is equal to the charge volume; acid strength is maintained at about 85-90%
Requires vigorous mixing to prevent polymerization reactions	Requires less energy in mixing due to lower viscosity of HF and increased temperatures
Products are fractionated in the deisobutanizer and debutanizer to separate isobutane & butane from alkylate; HF produced alkylate is fractionated in a similar manner	Octane number of HF produced alkylate is slightly lower than sulfuric acid alkylate

K. COKING

This process is used to produce light gases, naphtha, distillate fuel, heavy fuel oil and petroleum coke by cracking heavy residual products such as atmospheric and vacuum resids. Both *delayed coking* and *fluid coking* processes are utilized.

1. Delayed Coking

The delayed coking process is a semi-batch-type process whereby residual products are heated to temperatures ranging from 900°F to 975°F (482.2°C to 523.9°C) in a furnace and then transferred into a pressurized coking drum. Pressures range from about 10 psi to 100 psi. The volatile, thermally cracked products, termed coker gas oil, are removed from the coking drum and the coke which forms deposits onto the sides of the coke drum. To maintain somewhat of a continuous operation, an alternating cycle of filling and emptying of different coking drums is employed.

Characteristics of feedstock quality, recycle ratio and drum pressure affect the coke yield. Highly aromatic feedstock contains more carbon per feed volume and typically produces a high coke yield. Heavy coker gas oil can be recycled back into the coker feedstock to help improve the coke yield. Also, increasing the coking drum pressure tends to increase the coke yield. Typically, a higher coke yield results in a reduced liquid product yield.

After residing in the coke drum for about 24 hours, the coke, referred to as "green coke," is removed from the drum. Removal involves depressurizing the drum, filling the drum with water to cool the coke, draining the water, boring a hole into the coke bed and injecting high pressure water jets into the coke to break it free from the coking drum. Once free, the coke and water are separated. This entire process can take up to 48 hours.

Coke which is low in sulfur and metal content is valued as a fuel, as a raw material for the manufacture of electrodes and in graphite production. To produce high purity coke, all traces of volatile matter must be removed from coke. A calcination process is utilized for this purpose. This process requires the coke to be heated to temperatures of 2,000°F or higher. The pure coke is valued as raw material for the manufacture of electrolytic cell anodes and as a pure carbon source.

2. Fluid Coking

When compared to the delayed coking process, higher yields of liquid products are typically produced by fluid coking. This continuous process utilizes a fluidized reaction zone of hot coke particles held in motion by steam. The coke particles are first heated in a burner to temperatures ranging from 1100°F to 1200°F (593.3°C to 648.9°C). The hot coke particles then are blown into the reactor by steam. The residual fuel is fed into the reactor and cracks on the hot surface of the fluidized coke particles.

Vaporized products pass through cyclones, out of the reactor and into a scrubbing unit. Coke particles pass downward to the base of the unit and are steam stripped to remove residual vapors. The coke then passes into the burner, is again heated and then recycled back into the fluidized reactor. Large particles of coke are removed from the process in the burner. The smaller particles are recycled. Coke yield ranges from approximately 5% to 25% or more. Naphtha yields can range from 15% to 25% and distillate can range from approximately 50% to 75%. Light gases are also produced.

3. Flexicoking

This modern process utilizes elevated temperatures to react coke with air and steam to yield CO, CO_2, H_2, N_2, H_2O, H_2S and minute residual solids. After removal of residual solids and H_2S, the remaining gas can be used in burners and processing furnaces.

L. FINISHING PROCESSES

A number of processes exist which can be utilized to improve the quality of refined and blended fuels. These processes are often an essential part of ensuring that finished fuels are free of contamination and meet required specifications. Some of the common fuel constituents which can cause fuel storage, handling and performance problems are outlined in **TABLE 2-6.**

Processes such as caustic washing, sweetening, hydrodesulfurization and water removal are common examples of processes used to improve the quality of finished fuels. These processes are described in the remaining portion of this chapter.

1. Caustic Extraction and Caustic Washing

Aqueous solutions of NaOH or KOH are very effective at removing a variety of compounds from fuel. Components removed by extracting or washing the fuel with either of these caustic solutions include mercaptans, phenols, hydrogen sulfide and naphthenic acids. Removal of these compounds can improve the color, odor, stability, demulsibility and corrosiveness of fuel.

Thermal and catalytically cracked gasoline fractions can contain significant concentrations of phenols, low molecular weight organic acids and alkyl and aryl mercaptans. All of these compounds can initiate gum formation in gasoline. Caustic treatment readily removes these compounds.

Naphthenic acids occur primarily in distillate and some heavy fuel fractions. Typically, caustic treatment effectively removes these compounds. However, even after caustic treatment, alkali salts of heavier naphthenic acids may still remain oil soluble. In fuel, these compounds can act as very effective emulsifying agents.

Fuel haze and particulate contamination can be due to these acid salts. Caustic solutions of various strengths can be used to wash fuel. Usually 10-20 vol% of a 5-10% caustic wash solution is effective for most applications.

TABLE 2-6. Common Fuel Contaminants Resulting from Refining Processes

Fuel Contaminant	Source of Contaminant
Ammonia	Results from catalytic cracking and hydrotreating of nitrogen containing compounds, typically heterocyclic compounds
Aromatics	Natural components of crude oil; formed during cracking processes
Asphaltenes	Usually constituents of heavy fuel fractions; can form during cracking processes
Carbon Dioxide	Typically formed during catalytic cracking processes
Color Bodies	Sulfur, nitrogen, phenolic and conjugated compounds formed during processing; the result of oxidation and polymerization
Fatty Acids	Result of thermal cracking processes
Gums	Conjugated diolefins and other olefinic compounds formed during catalytic and thermal cracking processes; heterocyclic compounds present in fuel can also initiate gum formation
Hydrogen Sulfide and Mercaptans	Sour crudes; formed by decomposition of sulfur compounds during distillation, cracking, reforming and hydroprocessing
Naphthenic and Phenolic Acids	Certain crude oils; formed during cracking processes
Water	Crude oil; water washing, caustic washing; condensate water and contamination during storage

2. Sweetening Processes

Removal of the objectionable odors due to the presence of H_2S and mercaptans is the objective of the fuel sweetening process. Several methods can be utilized to remove these undesirable compounds including caustic washing, copper chloride sweetening, sulfuric acid treating, Merox processing, hydrotreating and other processes. These methods will be discussed below:

a. Fixed-Bed Copper Chloride Sweetening

The sour feed is first caustic washed to remove H_2S and then filtered through sand and rock salt to remove caustic and water. The feed containing mercaptan is then heated mildly and air or oxygen is sparged through the liquid. The oxygen-rich feed is passed upward through a fixed-bed reactor containing $CuCl_2$. The following series of reactions occurs:

$$CuCl_2 + 2RSH \rightarrow Cu(SR)_2 + 2HCl$$

$$Cu(SR)_2 + CuCl_2 \rightarrow Cu_2Cl_2 + RS\text{-}SR$$

$$Cu_2Cl_2 + 2HCl \rightarrow Cu_2Cl_2 \bullet 2HCl + 0.5\,O_2 \rightarrow 2CuCl_2 + H_2O$$

The disulfide which forms from the reaction does not have the objectionable odor of the mercaptan and does not aggressively corrode metals. The water which forms is removed from the process continuously.

b. Merox Process

This process is used frequently to sweeten gasoline and light distillate streams. In the first phase of Merox processing, mercaptans are extracted from the fuel with caustic. A sweetening process follows and involves sparging a caustic/fuel mixture with air in the presence of a catalyst. Remaining mercaptans then react to form disulfides which are noncorrosive and fuel soluble. Various catalysts such as metal chelates and phenylenediamine compounds are frequently utilized to improve the process effectiveness.

A combined Merox process operation is also utilized to sweeten fuel. This countercurrent method involves feeding fuel into the bottom of a vertical reactor and feeding caustic and catalyst from the top. Mercaptans are extracted into the caustic and react to form disulfides. The caustic and disulfides are later separated. The following reactions occur during the Merox sweetening process:

$$RSH + NaOH \rightarrow NaSR + H_2O$$

$$2\,NaSR + 0.5\,O_2 + H_2O \rightarrow RSSR + 2\,NaOH$$

c. Sulfuric Acid Treating

Sulfuric acid was used extensively several decades ago to reduce fuel mercaptans, olefins and to some extent, aromatics. Today, sulfuric acid processing has been almost entirely replaced by more effective and safer processes. Mercaptan removal was accomplished by converting mercaptans to disulfides through the following reaction:

$$RSH + H_2SO_4 \rightarrow RS\text{-}SO_3H + H_2O$$

$$RS\text{-}SO_3H + RSH \rightarrow (RS)_2SO_2 + H_2O$$

$$(RS)_2SO_2 \rightarrow R_2S_2 + SO_2$$

By mixing fuel with concentrated sulfuric acid at a ratio of about 99:1 held at a temperature between 90°F and 130°F (32.2°C and 54.4°C), about 90% of the mercaptans could be converted to disulfides.

This process was also effective at removing olefins by acting as a catalyst to initiate the polymerization of olefins into higher molecular weight compounds. During this process, some aromatics and color bodies were removed through reaction with olefins. Many of these polymers remained oil soluble.

3. Water and Particulate Removal

Water and particulates are the source of fuel quality problems such as filter plugging, corrosion and system component fouling. Water can be removed through salt drying, coalescence, filtration and good housekeeping.

a. Salt Drying

This process is typically used to remove water from distillate fuel and is effective at reducing the final concentration of water to about 75-100 ppm. During salt drying, fuel is passed up through a column or drum packed with sodium chloride or rock salt. Free water is removed using sodium chloride, but dispersed or emulsified water is usually not removed. If fuel contains high levels of dissolved water, calcium chloride must be used as a desiccant.

b. Coalescence

Electrostatic coalescers can be utilized for removal of free water from distillates and some solvents. During coalescence, fuel is heated to a temperature of about 100°F (37.8°C) and passed upward through a column containing electrodes and a desiccant. Free water coalesces onto the electrodes and the dried fuel then passes through a desiccant for additional water removal.

c. Filtration

Fuel filtration through sand is sometimes used to physically remove water and particulate matter from fuel. As fuel passes downward through a filtration drum, water and particulates are removed. Water haze can be removed if fuel passes upward through a filtration drum. In severe cases, a 5-micron filter may be required to produce bright and clear fuel.

d. Good Housekeeping Practices

The presence of condensate water and oxygen in a fuel storage tank can lead to problems with finished fuel quality. Microbial growth at the oil-water interface and the accumulation of rust and corrosion products in tank water bottoms can develop in water contaminated storage tanks. The turbulence which results during filling, blending and pumping operations can cause these contaminants to mix with fuel. Hazy fuel can result.

Regular removal of these water bottoms from contaminated storage tanks can help ensure that finished fuel remains bright and clear.

4. Molecular Sieves

Molecular sieves are used in a variety of fuel processing applications. Uses include drying and water removal from fuel, product purification, hydrocarbon separation and catalysis. Molecular sieves are composed of sodium and calcium aluminosilicate crystals which have been produced from natural or synthetic zeolite compounds. The crystals are dehydrated through heating and are processed to ensure that pore sizes are tightly controlled.

Molecular sieves function by both absorption and separation mechanisms. Through absorption, polar compounds such as water and H_2S are absorbed. Through separation, molecules which are held within the pores of the sieve are retained. Pore sizes are typically 3, 4, 5 and 10 Angstroms in diameter. Common uses for each of the pore sizes are included in **TABLE 2-7.**

TABLE 2-7. Uses of Molecular Sieves in Fuel Processing

Pore Diameter, Angstrom	Common Application
3	Water removal from olefins, methyl & ethyl alcohols and refrigerants
4	Water removal
5	Sweetening of light hydrocarbon streams; n-paraffin separation; improving jet fuel quality; purification of aromatics
10	Catalysis and filtration

The various pore size beads are also produced in different sphere diameters. Larger diameter beads are utilized primarily in liquid phase hydrocarbon separation processes, and smaller diameter beads are used primarily in vapor phase and certain liquid phase processes.

CHAPTER 3

Critical Properties of Crude Oil and Common Hydrocarbon Fuels

A. CRUDE OIL

Crude oil is composed primarily of hydrocarbon compounds. Organic and inorganic sulfur, oxygen and nitrogen containing species are also found in crude oil. Additionally, water, vanadium, nickel, sodium and other metals may be present.

The hydrocarbons present are complex, but are principally paraffinic, naphthenic and aromatic in composition. The complexity of the compounds present increases with the boiling range of the crude oil.

No generally accepted method of classifying crude oil exists. What is termed to be the ultimate crude oil analysis describes the composition as a percentage of carbon, hydrogen, nitrogen, oxygen and sulfur. Also, a chemical analysis is performed which describes the composition as a percent of paraffinic, naphthenic and aromatic type compounds.

These analyses are valuable in order to formulate a general idea of the usefulness of the crude oil in producing various refined products. However, the results do not yield much information about the variety of final products which can be obtained. To gain this information, an additional evaluation termed the "crude oil assay" is required.

1. Crude Oil Assay

A crude oil assay is based on a number of standard tests. Some of the more important tests are listed as follows:

a. Fractional Distillation

Fractional distillation characterizes the initial boiling point, volume % of materials boiling at specified temperature ranges, end point, residue and loss. A procedure termed 15:5 distillation is frequently used to assist in crude oil characterization. This procedure utilizes a column possessing the equivalent of 15 theoretical plates and a reflux ratio of 5:1.

b. Gravity

Either °API or specific gravity provides information on the paraffinic or aromatic nature of crude. The API gravity of most crude oil falls between the range of 20°API to 45°API.

c. Sulfur Content

High sulfur crudes, especially those with a sulfur content >0.5 wt%, complicate refining and are also more expensive to refine. Sulfur and some sulfur compounds are corrosive toward metals and can decrease the pH of the crude oil. The sulfur content of most crude oil varies from <0.1% to >5%.

The terms "sweet" and "sour" are often used to characterize crude oils. These terms were originally coined to identify the characteristic hydrogen sulfide odor of the crude oil. They are now commonly used to distinguish the crude oil sulfur level. In general, those crudes containing approximately 0.5 wt% sulfur or less are considered sweet, those containing greater than 0.5 wt% sulfur are considered "sour."

d. Nitrogen Content

Nitrogen is a severe refining catalyst poison. Crude oils containing >0.25 wt% nitrogen are difficult to refine.

e. Salt Content

Salts form scale which can foul refining equipment. Under refining conditions, salts can break down to liberate acids during processing. If the NaCl expressed salt content of crude oil is greater than 30 ppm, the crude oil must be desalted before processing.

f. Water and Sediment

Water and salts can lead to major problems such as equipment corrosion, uneven heating and equipment fouling.

g. Viscosity and Pour Point

The pour point can be important for establishing proper pumping, transportation and storage parameters.

h. Metals

The metal content can range from only a few parts per million to >1000 ppm. Trace elements such as iron, sodium, nickel, vanadium, lead and arsenic can corrode metallic parts and damage heating equipment. Low levels of nickel, vanadium and copper are known to deactivate refining catalysts.

2. Hydrocarbons in Crude Oil

Linear, branched and cyclic hydrocarbons are all found in crude oil. The concentrations, however, vary from crude source to crude source and even from oil well to oil well. The physical and chemical properties of the crude oil hydrocarbons can influence the way refineries process the crude and can also impact the performance of the refined fuels. A brief description of the properties of various crude oil hydrocarbon types is presented below:

a. Alkanes or Paraffins

The carbon number of crude oil paraffins typically ranges from C_1 to C_{40}. Some higher carbon number paraffin species have been identified, but their percentage is not large. Straight chain n-alkanes or linear paraffins of C_{20} or higher are responsible for wax related fuel problems such as high cloud points and high pour points.

Branched paraffins are commonly found in crude oil. Compounds with one or more methyl groups, ethyl groups and isopropyl groups have been identified. Ten carbon, branched paraffins are especially common.

b. Cycloalkanes or Cycloparaffins

These ring-like, saturated compounds are commonly called *naphthenes*. Often, about one-half of the total weight of the compounds found in crude oil are naphthenes. Compounds such as methyl derivatives of cyclopentane and cyclohexane are quite abundant in most crude oils. Single-ring naphthenics possessing either 5 or 6 carbon atoms in the ring structure are the most common. Higher or lower carbon number rings are less prevalent.

Methylcyclopentane   Methylcyclohexane

Bicyclic naphthenes are common in kerosene and diesel fuel. Both bicyclic and polycyclic naphthenes comprise a large portion of the mass of heavy distillates and residual oils. These higher molecular weight naphthenes possess four and five rings similar in structure to naturally occurring steroids and higher molecular weight terpenoids. They are also present in naphthenic lubricating oil fractions.

c. Alkenes or Olefins

Alkenes and cycloalkenes are commonly termed *olefins* in the refining industry. These double bonded compounds are rarely found in crude oil. Olefins appear in finished products as a result of cracking and other refining processes.

d. Arenes or Aromatics

Aromatic compounds are typically no more than fifteen percent of the total weight of most crude oil. The branched or alkylated aromatic derivatives are more prevalent than the non-alkylated aromatics. For example, toluene and xylene are found in higher percentages than benzene. Also, monoaromatic compounds such as toluene and xylene are some of the most common of the aromatics in crude.

Polynuclear aromatics can too be found. Naphthalene compounds, possessing two aromatic rings and phenanthrene compounds, possessing three aromatic rings are among the most common. Ring structures possessing four or more aromatic rings are less prevalent. Residual fuels, heavy gas oil fractions and lubricating oil raffinates can all contain polycyclic aromatics.

Biphenyl
(Example of a non-condensed or isolated ring structure)

Naphthalene
(Example of a condensed ring structure)

Higher molecular weight polynuclear aromatics can cause problems during processing of crude. They are known to contribute to deposit formation and fouling of refining equipment and fuel combustion furnaces. Also, some polynuclear aromatic compounds found in crude oil have been determined to be human carcinogens.

e. Naphthenic-Aromatics or Mixed Polycyclic Hydrocarbons

Naphthenic-aromatic compounds are formed by the condensation of an aromatic ring with a cycloparaffin. Examples of naphthenic-aromatics include indane, tetralin and their derivatives. These compounds are common constituents of distillate fuel and light gas oil fractions.

Indane

Tetralin

3. Other Compounds in Crude Oil

A variety of non-hydrocarbon species is found in crude oil. These compounds are found in all molecular weight ranges of crude oil components, but seem to concentrate in the heavier distillate and residual oil fractions. The effect of these materials on processing equipment, refining catalysts and finished product quality can be dramatic. Corrosion, catalyst poisoning and fuel stability problems can all be due to the effect of these non-hydrocarbon species.

a. Asphaltenes, Maltenes and Resins

Very high molecular weight, high boiling compounds composed of fused ring and waxy paraffin structures found in crude oil are called asphaltenes, maltenes and resins. Asphaltenes appear to exist within a colloidal dispersion in crude oil and are surrounded by resins and maltenes. The asphaltenes are amorphous, dark colored solids. Maltenes contain resins and higher molecular weight naphthenic and paraffinic waxes and oils. Resins are aromatic components of maltenes and are typically found in higher concentrations than asphaltenes. They are lighter in color and appear as viscous fluids rather than amorphous solids. Also, maltenes and resins are lower in molecular weight than asphaltenes and possess few condensed aromatic rings.

Asphaltenes are present as sheet-like structures composed of as many as twenty condensed aromatic rings. These rings can contain nitrogen, sulfur or oxygen heteroatom species and will typically possess naphthenic and paraffinic side chains. Molecular weights of individual asphaltene compounds can range from approximately 3,000 to 10,000. In micellar form, asphaltene molecular weight values can be significantly higher. Also, it is not unusual for heavy metals found in crude oil to be complexed with asphaltenes.

Asphaltenes can exist in various structural forms such as peri-condensed and kati-condensed rings. Upon heating, peri-condensed asphaltenes in crude oil have been shown to rearrange into kati-condensed asphaltenes.

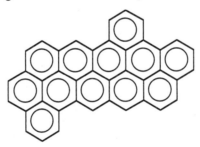

Peri-condensation of Aromatic Rings in Asphaltenes

Kati-condensation of Aromatic Rings in Asphaltenes

A significant portion of the sulfur and nitrogen containing species in crude oil can be found in heterocyclic form within the asphaltene, maltene and resin compounds. Oxygen containing heterocycles may also be present. Examples of high molecular weight aromatic, resinous and polar compounds found in crude oil are provided in **TABLE 3-1**.

TABLE 3-1. Examples of High Molecular Weight Aromatic, Resinous and Polar Compounds Found in Crude Oil

Compound
Aromatic amines
Benzo anthracenes
Benzo fluoranthracenes
Benzo pyrenes
Benzoic acids
Carbazoles
Chrysenes
Dibenzo thiophenes
Fluorenes
Imidazoles
Indoles
Naphthalenes
Naphtheno benzothiophenes
Phenanthrenes
Phenols
Pyrenes
Quinolines

b. Organosulfur Compounds

Following carbon and hydrogen, sulfur is the third most abundant element found in crude oil. The average sulfur content of crude oil is approximately 6500 ppm but can be identified at concentrations >5 wt% or more.

Sulfur can be found in most all petroleum fractions. In naphtha, distillates and some lube fractions, sulfur is a component of carbon and hydrogen containing molecules. In heavier fractions, sulfur may be bound in asphaltene and resin matrices in combination with nitrogen and oxygen.

The organosulfur compounds present in most fuels are either *thiols, sulfides* or *thiophenes.*

Thiols or Mercaptans

Also known as mercaptans, thiols possess very unpleasant odors and are most commonly found in the naphtha fraction. These low boiling compounds typically contain from one to eight carbon atoms and can be linear, branched or cyclic. Most thiols are removed from crude oil and petroleum products through hydroprocessing or sweetening.

Cyclopentyl Mercaptan

Sulfides

Diethyl sulfide and diethyl disulfide are examples of sulfides found in crude oil. Sulfide compounds are low in concentration and in molecular weight.

$$CH_3CH_2 - S - S - CH_2CH_3$$
Diethyl Disulfide

Thiophenes

Thiophene as a compound is not found in high concentration in crude oil. However, benzothiophenes and other aromatic thiophenes are critical components of high sulfur crude oil. Also, thiophenes are present in relatively high concentrations in oils containing increased percentages of aromatics, asphaltenes and resins.

2-Ethyl-Thiophene   **2-Methyl-Benzothiophene**

c. Organonitrogen Compounds

The nitrogen content is relatively low in most crude oils, usually <2000 ppm. The average concentration is approximately 950 ppm. The nitrogen content of most crude oil fractions increases with increasing boiling point. For example, naphtha and kerosene contain little nitrogen. Higher boiling distillates and residual oils can contain higher concentrations of organonitrogen compounds.

Organonitrogen compounds are quite stable throughout most refining processes. Basic heteroatomic species such as pyridine and quinoline can be found as well as non-basic pyrrole and indole compounds.

Quinoline   **Indole**

The nitrogen containing porphyrin ring species can be found in most asphaltic crude oils. The porphyrin ring usually holds a metal ion such as nickel or vanadium. These metals are known as powerful poisons for refining catalysts.

d. Oxygen Containing Compounds

The oxygen content of crude oil is typically low and is usually in the form of *naphthenic acid* compounds. Asphaltic crudes can contain significantly more of these acids than paraffinic crude oils. Also, the concentration of oxygen increases with the boiling point of the crude oil fraction.

In naphtha and light distillate components, oxygen containing compounds appear as carboxylic acids and phenols. Most of these compounds concentrate in the kerosene, fuel oil and lighter lubricant fractions. Straight-run gasoline, heavy distillates and residual fuels usually contain few acids.

The naphthenic acids in crude oil are primarily monocarboxylic acids possessing an alkylated, cyclopentane single-ring structure. Fused ring, branched aliphatic and dicarboxylic acid compounds are also found in lower numbers. Most species contain ten carbon atoms, but twenty carbon atom species have been identified.

A Naphthenic Acid

e. Metals and Metal Complexes

Nickel and vanadium are the most abundant metals found in crude oil. Other metals, metalloids and non-metals including aluminum, arsenic, barium, calcium, chromium, copper, gold, iron, lead, magnesium, manganese, phosphorus, silicon, silver and titanium can also be identified.

These elements exist in a variety of forms and concentrations. Metal complexes with asphaltenes, porphyrins and other polar organics can stabilize metal solubility throughout the crude oil matrix. Concentrations of metals can range from <1 ppm to 1500 ppm or more.

B. AUTOMOTIVE GASOLINE

Automotive gasoline is essentially a blend of hydrocarbons derived from petroleum and is used to fuel internal combustion, spark-ignition engines. Gasoline contains hundreds of individual hydrocarbons which range from n-butane to C_{11} hydrocarbons such as methyl naphthalene.

1. Typical Gasoline Quality Criteria

The typical boiling range for gasoline is 85°F to 440°F (29.4°C to 226.7°C). Properties of commercial gasolines are influenced by both refinery practice and by the nature of the crude oils refined. Typically, gasoline contains low molecular weight, branched paraffins, naphthenes and aromatics. The olefin content of gasoline commonly ranges from 5% to 20% but can vary depending upon the nature and amount of the blendstock utilized.

A gasoline blend may be composed of mixtures of components such as straight run gasoline, FCC gasoline, alkylate, reformate, isomerate and oxygenates. Many of the typical components used in gasoline blending are listed in **TABLE 3-2**.

Gasoline quality may be determined from the following criteria:

a. Antiknock performance

This is a physical and chemical gasoline combustion phenomena related to engine design and operating conditions. It is used to rate gasoline into grades such as "regular" and "premium." Fuel aromatics and isoparaffins help to limit knocking.

Engine knock is measured by two ASTM methods, ASTM D-2699 and D-2700. Method ASTM D-2699 is identified as the *Research Octane Number (RON)* and method ASTM D-2700 is identified as the *Motor Octane Number (MON)*. The primary differences between these two methods are summarized in **TABLE 3-3**.

TABLE 3-2. Typical Components Used to Prepare Finished Gasoline Blends

Gasoline Component	Characteristics
Normal Butane	Blended into gasoline to increase the vapor pressure for improved startability
Light Straight Run Gasoline	Composed primarily of components >C_5 with a cut point typically between 180°F to 200°F (82.2°C to 93.3°C). It is separated from heavy straight run gasoline and is caustic washed, hydrotreated or sweetened. The octane number is sometimes improved through isomerization
Heavy Straight Run Gasoline	Contains compounds with a boiling range from approximately 180°F to 380°F (82.2°C to 193.3°C). This fraction is typically blended directly into gasoline or used as reformer feedstock.
Cracked Gasoline; FCC Gasoline	Composed of paraffinic, olefinic and aromatic compounds; branched compounds are present in a relatively high amount; typically has a higher RON than MON; high olefin content FCC gasoline can lead to gum formation and fuel color degradation
Hydrocracked Gasoline	Relatively olefin-free, low aromatic cracked gasoline blendstock
Alkylate Gasoline	Composed of C_7 and C_8 branched hydrocarbons. This is a high octane number blending stock with MON values ranging from approximately 87 to 95 and RON values from 94 to 100
Isomerate	Contains C_4 to C_6 branched paraffins; used to improve startability and warmup properties
Reformate	Composed of higher octane aromatics, branched paraffins and cycloparaffins; high boiling reformate compounds can improve power, but may also contribute to deposits and oil dilution
Oxygenate	Ethanol, MTBE, ETBE, TAME; these components improve the octane number and help lower the CO and exhaust HC emission level of gasoline

TABLE 3-3. Research Octane Number and Motor Octane Number Test Parameters

Test Parameter	RON	MON
Test Result Value	Usually higher than MON	Usually lower than RON
Engine Speed, rpm	600	900
Air Temperature	60°F to 125°F Air intake temperature will vary with barometric pressure; temperature can be as high as 139°F	100°F
Lube Oil Grade	SAE 30	SAE 30
Carburetor venturi diameter	9/16 in. (14.3 mm)	9/16 in. (14.3 mm) at an altitude of 0 - 1600 ft (1000 m)
		19/32 in. (15.1 mm) at an altitude of 1600 - 3300 ft (500 - 1000 m)
		3/4 in. (19.1 mm) at an altitude over 3300 ft (1000 m)
Usefulness	Predicts relative performance under low speed, mild knocking conditions; best correlates with engine run-on tendency of engine	Correlates with high speed, high temperature operating conditions

The reported octane number for a gasoline is actually a calculated average of the RON and MON. This value is called the **Antiknock Index** and is stated as follows:

$$\text{Antiknock Index} = \frac{RON + MON}{2}$$

Isooctane and **heptane** are utilized as primary reference fuels for octane number determinations. Isooctane has an octane rating of one-hundred, and n-heptane has an octane rate of zero. Therefore, if a fuel has a reported octane number of 90, its performance is equal to that of a fuel blend containing 90% isooctane and 10% n-heptane

The term **Octane Number Requirement (ONR)** is used to describe the octane number of the fuel which will provide initial knock or trace knock, under high speed and load conditions. Engines vary widely in their ONR. This is due to both engine design and environmental conditions. The following conditions can affect ONR:

ONR may increase if:
- The compression ratio is increased
- The coolant temperature increases
- Combustion chamber deposits are present
- The ambient temperature increases

ONR may be minimized if:
- The absolute humidity decreases
- The altitude increases

Whenever new engines burn fuel, a certain amount of carbonaceous material builds up on the walls of the combustion chamber and the piston crown. These deposits act to increase the compression ratio and temperature of the cylinder environment. After about 10,000 miles, most vehicles accumulate enough deposit so that their ONR increases.

In some circumstances, the octane number of the gasoline used in the engine must be increased to prevent knock caused by accumulated deposits. This phenomenon is called **Octane Requirement Increase** or **ORI**. Both prevention and reversal of ORI are difficult to achieve.

The term **sensitivity** is used to describe the difference between the RON and MON of a fuel. High sensitivity fuels have a greater difference between the RON and MON values; 12 numbers, for example. Low sensitivity fuels have RON and MON only a few numbers apart. The sensitivity value represents the effect changes in operating conditions will have on the knock properties of the fuel.

The hydrocarbon composition of the fuel can influence sensitivity. Fuels containing higher percentages of unsaturated and aromatic compounds typically have high sensitivity. Linear paraffins in fuels are less sensitive, while highly branched paraffins can be more sensitive. **TABLE 3-4** contains information on the sensitivity of various fuel components.

TABLE 3-4. Sensitivity of Various Gasoline Components

Fuel Component	RON	MON	Sensitivity
n-Pentane	62	62	0
Isopentane	92	90	2
Toluene	120	115	5
2,3,4-trimethylpentane	103	96	6
Cyclohexane	83	77	6
Ethylbenzene	107	98	9
Isopropylbenzene	113	99	14
2,4,4-trimethylpentene	106	86	20
1,4-cyclohexadiene	75	40	35

b. Volatility

Gasoline volatility is measured by the Reid Vapor Pressure method or the Dry Vapor Pressure method. It is used to predict the cold startability of gasoline. If fuel volatility is too low, gasoline engines are difficult to start. Also, if the 10% distillation temperature is high, starting and warmup will both be impaired.

Reid vapor pressure and octane numbers for common gasoline blending components are provided in **TABLE 3-5**.

TABLE 3-5. Typical Vapor Pressure and Octane Number Values of Common Gasoline Blending Streams

Component	RVP, psi	MON	RON
n-Butane	59	89	93
Isopentane	20	87	92
n-Pentane	18	77	80
Light Straight Run or Virgin Naphtha	8	74	78
Light Catalytically Cracked Naphtha	4	79	93
Heavy Catalytically Cracked Naphtha	3	81	93
Light Alkylate	3	93	95
Heavy Alkylate	1	90	90
Reformate	1	93	101

c. Phosphorus Content

Gasoline must not contain phosphorus because it can degrade the activity of the catalytic converter catalyst.

d. Sulfur Content

The corrosion of metals is enhanced by fuel sulfur; also air polluting sulfur oxides form upon combustion of fuel sulfur. Sulfur ratings >0.15 wt% are considered high.

e. Existent Gum and Stability

These values predict the tendency of stored gasoline to form undesirable oxidation products and residue. Gums and residue may deposit on engine parts such as the intake manifold, intake valves and intake port. Poor engine performance may result.

f. Gravity

The °API is usually between °API 48 and °API 69.

g. Rust and Corrosion

Internal corrosion in storage tanks, transfer lines and underground pipelines can occur if fuel attacks and corrodes metal. Also, deposit formation and engine wear can result.

h. Hydrocarbon Composition

This is the breakdown of the percentage of paraffins, olefins, naphthenes and aromatics in gasoline. Fuel olefin content is closely monitored. High olefin containing gasoline is unstable and can lead to various fuel problems such as deposit formation and color degradation.

i. Additives

See **TABLE 3-6** and **TABLE 3-7** for a listing of the typical commercial gasoline additives. The following EPA regulations pertain to fuel additives:

- A fuel additive must contain only carbon, hydrogen, oxygen, nitrogen and sulfur. EPA approval is required for variance from this limitation.

- No more than 0.25 wt% of a fuel additive may be used. The additive must not contribute more than 15 ppm sulfur to the fuel.

- Up to 0.30 vol% methanol may be used; no more than 2.75 vol% methanol with an equal volume of butanol or higher molecular weight alcohol may be used.
- The fuel must not contain more than 2.0 wt% oxygen except fuels containing aliphatic ethers and alcohols other than methanol. These fuels can contain not more than 2.7 wt% oxygen. Higher percentages can be used if an EPA waiver is granted.
- Other exceptions must be reviewed and approved by the EPA.

TABLE 3-6. Typical Additives for Automotive Gasoline

Additive	Function
Antioxidant	Helps prevent oxidative degradation and gum formation in fuel
Corrosion Inhibitor	Two inhibitors could be used: a) Copper corrosion inhibitor to help prevent sulfur, hydrogen sulfide, and mercaptan attack on copper; and/or b) Ferrous metal corrosion inhibitor to prevent water/oxygen initiated corrosion of iron and steel system components
Metal Deactivator	Chelate metals which may act to catalyze fuel oxidation reactions
Demulsifier	Usually added to help prevent the fuel detergent additive from emulsifying with water; emulsified detergents will not perform
Detergents/Deposit Control Additives	Help prevent deposit buildup on fuel system parts such as carburetors, fuel injectors, intake valves, valve seats and valve guides; some additives will remove existing deposits
Dyes	Used to identify gasoline grades and types
Knock Control Compounds	Help control predetonation and ignition delay
Oxygenates	See **TABLE 3-7**

TABLE 3-7. Properties of Common Fuel Oxygenates

	MTBE	ETBE	TAME	Ethanol
R+M/2 Octane No.	108	112	105	115
RVP @ 100°F, PSI	7.8	4.0	2.5	2.3
IPB, °F	131	161	187	173
Oxygen Content, wt%	18.2	15.7	15.7	34.8
Solubility in Water, wt%	4.3	1.2	0.2	Infinite Solubility

Gasohol

When denatured ethanol is blended into gasoline at 10% by volume, the blend is identified as *gasohol*. Some of the important properties of gasohol compared to gasoline without 10% ethanol are included in **TABLE 3-8**.

TABLE 3-8. Effect of Ethanol Addition on Gasoline Properties

Property	Gasohol Effect
Antiknock Index	Normally increases up to three numbers
Reid Vapor Pressure	Increase about 0.5 to 1 psi
Copper Corrosion	Typically no change
Induction Time	Typically no change
Existent Gum	Typically no change
Water Tolerance	Phase separation can occur if fuel contains over 0.9 wt% water. Water tolerance can be improved by the addition of higher molecular weight alcohols such as butyl alcohol.
Metal Corrosion	Corrosive toward ferrous metal, magnesium, aluminum, terne plate and zinc alloys
Plastics and Elastomers	Degrades chlorinated polyethylene, neoprene, nitrile rubber, polysulfide, polyurethane
Paint	Polyurethane, alkyd and acrylic paints may be damaged after prolonged contact
Existing Deposits	Gums and adherent deposits can be removed from fuel storage and distribution system components
Hygroscopic Nature	Water will be absorbed by gasohol and may phase separate. Also, some water identifying pastes will not function in gasohol

2. Exhaust Emissions and Gasoline Detergents

There has been much discussion and a tremendous amount of time, money and energy put into designing engines and fuels which perform more efficiently and with lower emissions. Concern over fuel efficiency of operating engines and the level of hydrocarbons, CO, SO_x, NO_x and other compounds in gasoline exhaust coupled with the concern over the fugitive emissions from stored gasoline has brought about new regulations and new technology.

A significant reduction in fuel exhaust and fugitive emissions has resulted from new engine design and fuel technology. Presently, to help ensure that engines maintain a minimum level of cleanliness and combustion efficiency, PFI/IVD detergent use is mandated for use in U.S. gasolines. The additives approved for use and their required addition to fuel are both detailed by the Clean Air Act.

Engine tests such as ASTM D-5500, an IVD test using a BMW 4-cylinder, automatic transmission, fuel injected engine and ASTM D-5598, a PFI cleanliness test using a Chrysler 2.2 L turbocharged engine are both used to qualify detergent additives for performance. A Ford 2.3 L 4-cylinder engine test is also being considered for use to qualify IVD additives.

Reformulated gasolines containing controlled levels of sulfur, aromatics and olefins are now available, especially in regions having high amounts of atmospheric ozone. Also, oxygenates such as MTBE and ETBE are added at varying levels to reformulated gasolines. When combusted, these fuels burn with lower SO_x, CO and HC emissions than fuels produced in previous decades. Fuel vapor pressure limits are also established to help minimize fugitive emissions from storage tanks and from fueling nozzles.

3. Combustion Chamber Deposits

Deposits can form within the combustion chamber of a gasoline engine. These deposits are complex in nature and can contain components from fuel combustion, from the lubricating oil and from some additives. Typically, the heavier fuel and lubricant components such as condensed aromatics and bright stocks contribute most significantly to combustion chamber deposits.

Once formed, the deposits can prevent adequate heat transfer from the combustion chamber. On the metal surfaces of the combustion chamber, the deposits act as an insulator. The temperature within the chamber can then increase to promote preignition and knock. If deposits continue to build, they can lead to an increase in the compression ratio within the combustion chamber and promote knock.

C. AVIATION GASOLINE

Aviation gasoline is possibly the most complex fuel produced in the refinery. Specifications for volatility, calorific value and antiknock ratings are stringent. Aviation gasoline contains a high concentration of alkylate and high octane number aromatics. Careful fractionation of virgin naphtha can also be performed to isolate isohexane and isoheptane fractions. Aromatic components are limited by their calorific value and are typically C_7 and C_8.

1. Typical Aviation Gasoline Quality Criteria

a. Antiknock Properties

Various fuel grades are designed to guarantee knock-free operation. Knocking combustion can damage the engine and can result in serious power loss if allowed to persist.

Lean Rating This rating is termed the Aviation Octane Rating (F-3) and is used to simulate lean fuel/air conditions experienced during aircraft cruise operation. The number is determined by ASTM D-2700 MON procedure and converted to the Aviation Rating.

Rich Rating This rating is termed the Supercharged Octane Rating (F-4) and is used to simulate takeoff conditions. The rating obtained is typically higher than the lean rating.

Performance Number This value is primarily used to rate the antiknock values of aviation gasolines with octane numbers over 100. It is defined as the maximum knock-free power output obtained from fuel expressed as a percentage of the power obtainable from isooctane. The relationship between octane number and performance number is listed as follows:

$$\text{Octane Number} = 100 + \frac{\text{Performance Number} - 100}{3}$$

b. Volatility

Both low end and high end volatility limits are established. Low end volatility control ensures that fuel will vaporize adequately for engine starting and distribution throughout the combustion system. Upper end volatility limits ensure that "vapor locking" or fuel line icing will not occur. Isopentane and certain C_5 fractions are used to help control vapor pressure.

c. Freezing Point

This value is important to help ensure that wax crystals do not form and interfere with the flow and filtration of fuel at high altitudes. At an altitude of 25,000 feet, air temperature of -60°F to -70°F (-51.1°C to -56.7°C) is possible.

d. Storage Stability

Gums which form in aviation gasoline could produce undesirable deposits on induction manifolds, carburetors, valves, injectors, etc. Strict limitations exist for the gum content of aviation gasoline.

e. Sulfur Content

Sulfur is limited to 0.05 wt%. Sulfur degrades the efficiency of antiknock compounds and can lead to metal corrosion.

f. Water Reaction

The original intent of this specification was to prevent the addition of high octane, water soluble components such as alcohol or diisopropyl ether to aviation gasoline. Present specifications require fuel, water and interface ratings by ASTM D-1094.

A listing of typical specification properties for aviation gasoline is provided in **TABLE 3-9.**

TABLE 3-9. Selected Typical Properties of Aviation Gasoline Grades from ASTM D-910 Standard Specification for Aviation Gasoline

Property	Grade 80	Grade 100	Grade 100LL
Knock Value, min. Octane Number, Lean Rating	80	100	100
Knock Value, min. Rich Rating: Minimum Octane Number	87	—	—
Knock Value, min. Rich Rating: Performance Number	—	130	130
Freeze Point, °C (°F)	-58 (-72)	-58 (-72)	-58 (-72)
Electrical Conductivity, pS	450	450	450
Net Heat of Combustion, Min, BTU/lb	18,720	18,720	18,720
Vapor Pressure, kPa (psi)	38 (5.5) min. 49 (7.0) max.	38 (5.5) min. 49 (7.0) max.	38 (5.5) min. 49 (7.0) max.
Sulfur, max, %m	0.05	0.05	0.05
Tetraethyl lead, max, ml/U.S. gal	0.5	4.0	2.0
Color	Red	Green	Blue

D. JET FUEL OR AVIATION TURBINE FUEL

Grades of commercial and military jet fuel are virtually identical worldwide and differ mainly in the types of additives permitted. The only significant exception is in the type of fuel used in Russia and many East European countries.

Jet fuels are typically prepared from either straight-run kerosene or from wide-cut kerosene/naphtha blends off of the atmospheric distillation unit. **TABLE 3-10** briefly describes the composition of some typical jet fuel grades:

TABLE 3-10. Aviation Fuel Grades and Characteristics

Grade		Characteristics
U.S.	NATO	
JP-4	F-40	No longer produced; wide cut naphtha based; Air Force standard
JP-5	F-44	High flash/high freeze point kerosene based; Naval carrier aircraft fuel; will not corrode copper and will not emulsify with salt water
JP-7	—	Low-volatility kerosene; very high performance aircraft fuel
JP-8	F-34	Kerosene (Jet A-1 type); Air Force standard turbine fuel; replaces JP-4
—	F-35	JP-8/F-34 without corrosion inhibitor/lubricity improver or icing inhibitor; very similar to ASTM D-1655 Jet A-1, commercial fuel
Jet A /Jet A-1		Jet A freeze point -40°C; Jet A-1 freeze point -47°C

1. Typical Jet Fuel or Aviation Turbine Fuel Quality Criteria

a. Chemical Properties and Combustion

Jet fuels are blended primarily from straight run distillate components and contain virtually no olefins. Aromatics in jet fuel are also limited. High aromatic content can cause smoke to form during combustion and can lead to carbon deposition in engines. A total aromatic content >30% can cause deterioration of aircraft fuel system elastomers and lead to fuel leakage.

b. Sulfur Content

Jet fuel sulfur normally ranges from 0.20 wt% to 0.40 wt%. Metal corrosion may result when high sulfur fuel is burned. Mercaptan sulfur is limited to <50 ppm. Odor, elastomer degradation and corrosion toward metals result from high mercaptan levels.

c. Oxygenated Organics

Oxygen containing impurities such as phenols and naphthenic acids can adversely affect water separation properties and initiate gum formation. No limit presently exists to control the amount of oxidized organic compounds found in jet fuel. However, tests for existent gums, neutralization number and water separation indirectly limit the presence of oxygenated materials in jet fuel.

Fuel system icing inhibitors such as EGME and DEGME contain oxygen, but are added within controlled limits.

d. Volatility

The 10%, 20%, 50% and 90% distillation points are specified to ensure that a properly balanced fuel is produced. Vapor pressure is also related to cold engine starting and helps limit vapor lock at high altitudes.

e. Heat of Combustion

A high density fuel is preferred because it normally has the greatest heating value per unit volume of fuel. Jet fuels derived from paraffinic crudes have a slightly lower density but higher calorific value per equivalent mass than those refined from naphthenic crudes.

f. Low Temperature Properties

Fuel must flow during long periods at high altitudes. Fuel must not solidify and block lines, filters and nozzles. Low temperature viscosity must also be controlled to ensure that adequate fuel flow and pressures are maintained. Fuel viscosity can also significantly affect the fuel pump service life.

g. Combustion Quality

This property is based primarily on the paraffin, naphthene and aromatic (principally diaromatic) content of the jet fuel. The smoke point together with the naphthalene content help to characterize fuel combustion quality.

h. Thermal Stability

Fuel must not produce lacquer or form deposits which can adversely affect jet engine operation. The Jet Fuel Thermal Oxidation Test (JFTOT) helps predict the tendency of fuels to form deposits.

i. Contaminants

Clay filtration is commonly used at airports and terminals to remove water, solids and unwanted compounds prior to use of the jet fuel.

j. Water Retention and Separation

These determinations help to predict whether fuel viscosity, density and trace contaminants such as sulfonic and naphthenic acids work together to retain water and/or fine particulate matter in jet fuel. Very small traces of free water or particulates could result in ice formation and adversely affect jet engine operation.

k. Electrical Conductivity

The electrical conductivity of fuel must be maintained at a minimum and maximum level. Fuel must have adequate conductivity to ensure that static charge does not build up in the fuel. However, if fuel is too highly conductive, some capacitance-type aircraft fuel gauges can be disrupted.

l. Aviation Fuel Additives

Additives are introduced into aviation fuel for a variety of reasons. Primarily, additives are added to improve fuel performance or to prevent undesirable behavior. They may be introduced into fuel under the following limitations:

Mandatory The additive must be present within minimum and maximum limits defined.

Permitted The additive may be added by the fuel producer up to a maximum limit.

Optional The additive may be added with user/purchaser agreement.

Not Allowed The additives not listed are not permitted.

Approved Aviation Fuel Additives

Antioxidants - prevent gum formation in fuel. The approved additives are generally phenolic based. Use is mandatory in aviation gasoline. Antioxidants are permitted in civil and military jet fuels. Antioxidants are mandatory in hydroprocessed British and U.S. military jet fuels as well as international civil Jet A-1.

Metal Deactivator - chelates metal ions, primarily copper. Copper catalyzes the oxidation and degradation of jet fuel. Use is not permitted in aviation gasoline. A metal deactivator is permitted in civil and military jet fuels.

Corrosion Inhibitor - helps prevent rusting of metal engine components. Also, corrosion inhibitors provide a protective film on metal surfaces to aid in improving the lubricating properties of jet fuel. Use is not permitted in aviation gasoline and civil jet fuels, but is mandatory in military jet fuel grades.

Fuel System Icing Inhibitor - prevents ice crystal formation in jet fuel systems. Ethylene glycol monomethyl ether (EGME) and diethylene glycol monomethyl ether (DEGME) are the approved anti-icing additives for jet fuel use. Known by the U.S. military as Grade FSII. Not permitted in aviation gasoline and civil jet fuel unless special agreements are in place. FSII is mandatory in military jet fuel.

Anti-static Additive - dissipates static charge in jet fuel. Static charge buildup can result in unwanted ignition of jet fuel/air mixtures. Use is normally not permitted in aviation gasoline except in Canada and Britain. Use is permitted in civil jet fuel and mandatory in military jet fuel.

Anti-smoke Additives - all are ash containing and are not approved for flight use, only for engine testing.

Microbiocides - used to prevent microbial growth in jet fuel. The only product approved for use in jet fuel is *Biobor JF*. However, its use is permitted only infrequently. The maximum treat rate is 270 mg/L.

Lubricity Additive - used to help prevent wear of high pressure fuel injection equipment. Use is not permitted in aviation gasoline and can be used only by permission in civil jet fuel. Use is mandatory in military jet fuel grades. Typically, the corrosion inhibitor also functions in providing adequate fuel lubricity performance.

Tetraethyl Lead - used to improve the octane number (antiknock) properties of gasoline. Use is mandatory in aviation gasoline. Tetraethyl lead is not permitted in civil and military jet fuel.

A brief listing of typical specification properties for Jet A, Jet A-1 and Jet B is provided in **TABLE 3-11.**

TABLE 3-11. Selected Typical Physical Properties of Jet A, Jet A-1 and Jet B

Property	Jet A	Jet A-1	Jet B
Flash, °C (°F)	38 (100.4)	38 (100.4)	<-18 (<0)
Freezing Point, °C (°F)	-40 (-40)	-47 (-52.6)	-50 (-58)
Existent Gum, mg/100 ml	7	7	7
Color	Water White to Light Amber	Water White to Light Amber	Water White to Light Amber

E. DIESEL FUEL AND FUEL OIL

Fuels used to power marine, truck, automotive, railroad and industrial diesel engines fall into the general category of diesel fuel. Heating oil, burner fuel, kerosene and heavier residual fuels fall primarily into the category of fuel oil.

Today's diesel fuel grades are blends which may contain straight run distillate, cycle oil, various gas oils and heavy cracked distillates. Kerosene or jet fuel may be blended into the diesel to improve the low temperature viscosity and handling characteristics of the fuel.

Lighter fuel oil grades contain many of the same blend components as diesel fuel grades. Heavier fuel oils such as No. 5 and No. 6 contain vacuum gas oils and residual fractions.

In the U.S. and other parts of the world, both low sulfur diesel fuel and high sulfur diesel fuel are being refined. Because fuel sulfur level has been identified as the primary component of fuel emission particulates and "acid rain," sulfur reduction has been mandated and implemented.

Hydroprocessing is utilized to reduce fuel sulfur concentrations. Fuels classed as *high sulfur diesel* typically contain up to 5000 ppm of sulfur. Fuels classed as *low sulfur* diesel typically contain up to 500 ppm of sulfur. The sulfur removed by hydroprocessing is converted to H_2S. Occasionally, some light mercaptan carries into the fuel and is removed by washing or stripping processes.

In comparing high sulfur diesel fuel with low sulfur diesel fuel, the following major performance differences have been determined and are presented in **TABLE 3-12.**

TABLE 3-12. Comparison of High Sulfur Diesel and Low Sulfur Diesel Fuel

Test Parameter	High Sulfur/Low Sulfur Difference
Cetane Number	Low sulfur fuel engine cetane number is typically higher than high sulfur fuel. The range can be from 1 to 5
Cloud Point	Variable
Pour Point	Variable
CFPP/LTFT	Variable
Lubricity	Low sulfur fuel is typically 500 to 1000 g lower as measured by the "Scuffing BOCLE" test procedure
Ferrous Corrosion	Low sulfur fuel typically provides poorer protection against rusting than high sulfur fuel
Electrical Conductivity	Low sulfur fuel typically has a very poor electrical conductivity performance
Oxidative/Thermal Stability	Low sulfur fuel is typically significantly more stable to oxidative or thermal degradation. Peroxide formation is being monitored as an indication of instability in low sulfur fuels since conventional testing methods do not provide readily discernible differences

1. Typical Diesel Fuel and Fuel Oil Quality Criteria

a. Fuel Grade

Fuel grade is based primarily upon the viscosity of the fuel and the intended application. Both *Diesel Fuel* and *Fuel Oil* grades exist.

Diesel fuel grades and their intended applications are outlined in **TABLE 3-13**.

TABLE 3-13. ASTM D-975 Diesel Fuel Grades and Applications

ASTM D-975 Diesel Fuel Grade	Application
No. 1-D Low Sulfur	Very low pour point fuel for use in high speed engines requiring low sulfur fuel; also, low sulfur kerosene applications
No. 2-D Low Sulfur	High speed engines requiring low sulfur fuel
No. 1-D	Very low pour point fuel for use in high speed engines utilizing higher sulfur content fuel; also higher sulfur kerosene applications
No. 2-D	High speed engines utilizing higher sulfur content fuel
No. 4-D	High viscosity fuel for use in medium and low speed engines utilized in sustained load, constant speed applications

No. 1-D and No. 2-D are most commonly used in truck, railroad and in some stationary engines. Grade No. 4-D fuels are used in marine and certain industrial diesel applications.

Fuel oil grades and their intended applications are outlined in **TABLE 3-14.**

Critical Properties of Crude Oil and Common Hydrocarbon Fuels

TABLE 3-14. ASTM D-396 Fuel Oil Grades and Applications

ASTM D-396 Fuel Oil Grade	Application
No. 1	Vaporizing type burners
No. 2	Atomizing type domestic and industrial burners
No. 4 Light	Low pour point heavy distillate fuel for use in pressure-atomizing type burners
No. 4	Heavy distillate for use in pressure-atomizing type burners
No. 5 Light	Low pour point fuel for use in residual fuel oil burners
No. 5 Heavy	Residual fuel oil burners
No. 6 or Bunker C	High viscosity oil usually requiring heating before use in residual fuel oil burners

Federal diesel fuel oil classifications are provided in **TABLE 3-15**.

TABLE 3-15. Federal Specification VV-F-800 Describing Diesel Fuel Oil Classifications

Military Symbol	NATO Code	Description
DF-A	-	Arctic Grade
DF-1	-	Winter Grade
DF-2	F-54	Regular Grade

ARCTIC GRADE

Arctic grade diesel fuel oil is recommended for use in environments with ambient temperatures lower than -32°C (-25.6°F). It is intended for use in high speed automotive-type diesel engines, pot-type burner space heaters and non-aircraft turbine engines. It is not intended for use in slow speed stationary engine applications.

WINTER GRADE

Winter grade diesel fuel oil is recommended for use in ambient temperature environments as low as -32°C (-25.6°F). It is intended for use in high speed automotive type diesel engines and non-aircraft type turbine engines. This fuel may be used for medium-speed stationary engine applications.

REGULAR GRADE

Regular grade diesel fuel oil is recommended for use in temperate ambient temperature environments. It is recommended for use in all automotive high-speed and medium-speed engine applications and non-aircraft turbine engines.

b. Cetane Number

The cetane number is a measure of fuel combustibility using a variable compression ratio cetane engine. High cetane number fuels generally start easier in cold weather, provide more complete combustion and yield an overall increase in engine efficiency, primarily power.

c. Cetane Index

The cetane index is an estimation of the cetane number from fuel physical properties such as the 10%, 50% and 90% boiling points and the °API.

d. Gravity

Diesel fuels usually have °API values between °API 23 and °API 49.

e. Distillation

The distillation profile can be used to predict the following:
- *Startability* - poor if 10% point is too high
- *Warmup time* - increased if boiling range is wide between 10% and 50%
- *Smoke & Odor* - minimized if 50% point is low
- *Carbon residue* - increased if 90% point is high

f. Viscosity

The shape of the fuel spray is related to viscosity. High viscosities cause poor atomization and a solid stream jet spray pattern. Poor combustion and low power result. Low viscosities result in soft, non-penetrating fuel spray, leakage of fuel past the injection plunger and possible wear of fuel system components.

g. Heat of Combustion

The heat of combustion can be determined from the °API gravity. Low °API fuels have higher BTU ratings per pound than high °API gravity fuels. See **APPENDIX 1** for °API vs. BTU nomograph.

These values can be determined from the following equations:

CALCULATED CARBON AROMATICITY INDEX

$$CCAI = D - 140.7 \log\log(V + 0.85) - 80.6$$

where: D = Density in kg/m³ @ 15°C
V = Viscosity in cSt @ 50°C

CALCULATED IGNITION INDEX

$$CII = 275.985 - (0.254565)D + 23.708 \log\log(V + 0.7)$$

where: D = Density in kg/m³ @ 15°C
V = Viscosity in cSt @ 50°C

Bunker fuels with CCAI values <860 would be considered to be of better quality than fuels with values >860. Fuels with CII values <30 are considered to be of poor quality.

d. Density

Marine fuel density is reported in values of kg/m³ or g/ml. Typical densities range from about 0.900 to 0.990 g/ml. As fuel density approaches 1.000 g/ml, water separation properties will be impaired and ignition properties will be poorer than lower density fuel.

e. Sulfur Content

Marine fuel sulfur can range from 1.5 wt% for DMA to as high as 5.0% for RME and higher viscosity grades of marine residual fuel. Problems related to sulfur include high SO_x emissions and the formation of sulfuric and other acids within the fuel combustion system. At low temperatures, the formation and condensation of acids within the combustion chamber can result in corrosion and wear of metal system components.

f. Total Sediment

Total sediment includes measurement of insoluble fuel asphaltenes, inorganic compounds, catalyst fines and other compounds which can be filtered from the fuel. This value is important because sediment may plug filters and strainers, overload centrifuge systems and contribute to fuel system deposits.

An additional method identified as Sediment by Extraction involves the use of toluene as an extracting medium prior to determining the sediment value. This method may result in removal of some compounds such as asphaltenes from the final sediment determination.

g. Viscosity

Fuel viscosity directly influences the pumping and atomization characteristics of the fuel. High viscosity fuel may be difficult to efficiently pump through lines and filters unless it is heated. Also, high viscosity fuels may not atomize and finely disperse when injected into the combustion chamber of an engine or into the firebox of a boiler. Incomplete atomization results in poor fuel efficiency and high hydrocarbon emission values.

G. BURNER FUELS

Although LPG is used extensively as a burner fuel, #1 fuel oil, termed kerosene or range oil, as well as #2 fuel oil, often termed heating oil, are frequently used. These fuels must have several important properties to be considered as a good quality burner fuel. The qualities typically include good burning quality, high heat content, low smoking tendency, good low temperature handling characteristics, good ignitability and low deposit forming tendency.

Desired burner fuel performance characteristics and fuel properties which affect performance are provided in **TABLE 3-18**.

TABLE 3-17. Fuels Typically Used in Marine Applications

Fuel Type	Description
Marine Gas Oil (MGO) Grade DMA	Grade DMA is blended from straight run and cracked gas oils. This fuel is sometimes identical to #2 diesel fuel and is used in high speed engines.
Marine Diesel Oil (MDO) Grades DMB and DMC	Grade DMB is blended from gas oils and a trace amount of residual oil. It has a minimum cetane number of 35 and a maximum viscosity of 11 cSt @ 40°C.
	Grade DMC is blended from gas oil components and up to 10% residual oil. It has a minimum cetane number of 35 and a maximum viscosity of 14 cSt @ 40°C.
	Marine diesel oil is used in both high speed and medium speed engines.
Marine Intermediate Fuels	Various viscosity grades are blends of residual fuel and either MGO or MDO. Typical viscosity grades include IF 30 (30 cSt @ 50°C), IF 60 (60 cSt @ 50°C), IF 100 (100 cSt @ 50°C), IF 180 (180 cSt @ 50°C) and IF 380 (380 cSt @ 50°C). Grades IF 180 and IF 380 are the most common.
Marine Residual Fuels Bunker Fuel Oil	Grades ISO RMA thru RML marine residual fuel and bunker fuel are blended from components such as atmospheric resid, vacuum resid, visbreaker resid, FCC bottoms, low grade distillate and cracked components. Bunker fuel has a maximum viscosity of 550 cSt @ 50°C, density of 0.990 g/cc and sediment of 0.1 wt%. ISO marine fuel oil viscosities range from 10 cSt to 55 cSt @ 100°C. These fuels are used in slow speed diesel engines and boilers.
Naval Distillate Fuel NATO F-76	This fuel has specific physical and chemical requirements listed in specification MIL-F-16884J. The physical properties are similar to those required for MGO Grade DMA. Restrictions on additive addition are provided.

Speed Engines

These engines operate at over 1000 RPM and are found in small pleasure boats, boats, life boats and small portable power generators.

Medium Speed Engines

These engines operate at speeds of 500 to 1000 RPM and can develop 100-200 HP cylinder with cylinder bore diameters up to 12 inches. Small stationary and marine power plants can be medium speed engines. Also, railroad engines can be this type.

Low Speed Engines

These engines operate up to 500 RPM and can develop 1000 HP at cylinder speeds 80 to 150 RPM. Cylinder bore diameter measurements can be up to 36 inches. These engines are limited primarily to large stationary and marine power plants.

Examples of marine fuels are listed in **TABLE 3-17**.

Some of the more critical properties related to marine fuels include ash content, carbon residue, CCAI, density, sulfur, total sediment and viscosity. A description of these properties and the primary reason for their implementation are provided below:

a. Ash Content

Residual oils, FCC clarified oils and other heavy petroleum fractions used to blend marine bunker fuels may contain catalyst fines and other metals. When delivered with fuel, these metals may score pistons and piston liners to the extent that engine performance is seriously impacted. For this reason, an aluminum plus silicon specification of 25 mg/kg has been established for DMC fuels and a specification of 80 mg/kg has been established for marine residual fuels.

b. Carbon Residue

Fuel asphaltenes, resins and other heavy compounds can build up as residues on engine components after evaporation and burning away of the more volatile fuel components. These residues can accumulate as deposits which may interfere with heat transfer, lubrication and efficient fuel combustion.

c. Calculated Carbon Aromaticity Index (CCAI) and Calculated Ignition Index (CII)

It is impractical to determine the cetane number of residual fuels in the ASTM D-613 cetane engine. Because of this, the Calculated Carbon Aromaticity Index and the Calculated Ignition Index were respectively developed by Shell and BP.

o. Stability

Fuel stability is an indication of the sediment and gum forming tendency of fuel. Gums and sediment can cause filter plugging, combustion chamber deposits and can result in sticking of pumping and injection system components.

p. Water and Sediment

Water and sediment usually result from poor fuel handling and storage practices. Water and sediment can enter from atmospheric air, through fuel transportation systems and from blending operations. Corrosion, filter blocking and injection system wear and deposit formation can result.

A listing of some classes of compounds which can be found in middle distillate fuel is provided in **TABLE 3-16**.

TABLE 3-16. Representative Paraffin, Naphthene and Aromatic (PNA) Analysis of a Middle Distillate Fuel by Mass Spectroscopy

Fuel Component	LV %
Paraffins	32.30
Monocycloparaffins	13.48
Dicycloparaffins	9.03
Tricycloparaffins	2.81
Alkylbenzenes	6.23
Indanes/Tetralins	4.98
Indenes	2.58
Naphthalene	0.25
Naphthalenes	8.61
Acenaphthenes	8.02
Acenaphthalenes	6.74
Tricyclic Aromatics	4.97
TOTAL	100.00

F. MARINE FUELS

Fuels used in marine applications are quite diverse in their properties. Low viscosit distillate fuels and high viscosity residual fuels can both be considered marin fuels. The applications, though, would differ and could include use in direct injecte diesel engines, boilers and gas turbines. Also, high speed, medium speed and slo\ speed engines can be found in marine applications.

h. Cloud Point

The cloud point measurement is used to predict the temperature at which wax in fuel may begin causing operating problems such as filter plugging and blockage of lines in fuel systems.

i. Pour Point

The pour point value is used to predict the temperature at which cold fuel will gel and no longer flow.

j. Flash Point

This is an important fuel transportation, handling and storage property. It can also be used to help predict contamination with more volatile compounds such as gasoline.

k. Sulfur Level

There is much concern about the emissions which result when fuel sulfur combusts (i.e., sulfur oxides). These gaseous products further react to form environmental pollutants such as sulfuric acid and metal sulfates. Active sulfur and certain sulfur compounds can corrode injection systems and contribute to combustion chamber deposits. Under low temperature operating conditions, moisture can condense within the engine. Sulfur compounds can then combine with water to form corrosive acidic compounds.

l. Carbon Residue

This value helps predict the deposit forming tendency of fuel. Deposits in oil burner systems can form "hot spots" on surfaces which can lead to stress, distortion and even cracking of system components.

m. Ash Content

These are the nonburnable components, typically metals and metalloids, found in fuel. Depending upon size, these particles can contribute to fuel system wear and filter and nozzle plugging. Sodium, potassium, lead and vanadium can cause corrosion of certain high temperature alloys such as those found on diesel engine valves and gas turbine blades.

n. Neutralization Number

This value is a measure of the acidity or basicity of fuel. The presence of acidic compounds can indicate a fuel stability or oxidative degradation problem.

TABLE 3-18. Burner Fuel Performance Properties

Desired Performance	Fuel Property Affecting Performance
Ignites Easily and Safely	The flash point, initial boiling point and viscosity values affect ignitability. If the values are too high, the fuel may not ignite readily.
Good Flame Height	If the fuel viscosity is too high, it may not travel readily through a wick and burn effectively. High fuel viscosity may be a reflection of high IBP and/or EP temperatures. Also, water contamination can interfere with fuel burning quality.
Low Smoke and Soot	Fuel paraffins burn with less smoke than aromatic compounds and can provide a higher flame height without smoking. High fuel density indicates the presence of greater concentrations of fuel aromatics. Also, certain sulfur containing compounds can burn to form lamp deposits.
Low Deposit Forming Tendency	High boiling components do not volatilize and burn completely. High distillation residue, EP, and carbon number values may be linked to deposit formation tendency.
Good Fuel Economy and Heating Value	Higher density fuel has more potential heating value or BTU per volume than lower density fuel. High BTU fuel has a greater concentration of aromatic compounds per fuel volume than low BTU value fuel.
Low Odor	Low fuel sulfur content helps to minimize combustion vapor odor.
Remains Liquid at Low Temperatures	The pour point is an indication of the low temperature handling properties of the fuel. High density, low IBP, low EP, low viscosity fuels will typically have good low temperature handling properties.

The most common types of kerosene burning equipment typically include the following:

Wick Vaporizing Burner
This type of burner is found in lamps and portable stoves used for cooking and heating.

Perforated Sleeve Vaporizing Burner
This type of burner is used on ranges. Fuel burns with a silent, blue flame maintained by a natural air draft.

Vaporizing Pot Burner
The burners are used in room space heaters, water heaters and in central heating systems. Both natural air draft and forced air draft versions are available.

Common systems which utilize #2 heating oil include home heating and boiler systems. Typical atomizers used in these systems include the following:

High Pressure Atomizing Gun Burner
This burner utilizes a high pressure fuel atomizer to spray and meter heating oil into the firebox. Combustion air is supplied independently. An electronic ignition system is used to ignite the air fuel mixture.

Low Pressure Air Atomizing Burner
Both air and fuel are injected through the same atomizing nozzle in this burner. Fuel is injected at a low pressure while air is injected at a high velocity resulting in a finer spray than is provided by the high pressure atomizing gun. A secondary, external air supply is also provided. This type of atomizer is less susceptible to nozzle plugging.

H. RESIDUAL FUEL OIL

The terms "residual fuel" or sometimes "resid" are used to describe high boiling fractions obtained from crude oil distillation and processing. The following petroleum fractions are often described as residual oils:

- Atmospheric tower bottoms or reduced crude oil
- Visbreaker bottoms
- Heavy vacuum gas oil
- Vacuum tower bottoms
- FCC bottoms

These fractions can be used directly as refined or may be blended with lighter fractions to produce the following finished products:

- #4 fuel oil
- #5 light or heavy fuel oil
- #6 fuel oil
- Bunker fuel oil
- Marine diesel oil
- Marine intermediate fuels

Residual fuels are usually quite high in their BTU rating and are commonly used to fire industrial, commercial and home heating systems. They are used to heat water to produce steam for use in radiant heating systems and in power generation. Energy for the production of steam to power turbines often comes from residual fuel fired systems. They are also used to fuel direct injected marine diesel engines and other medium and low speed engines.

The high flash point, good storage stability and potential energy value make residual fuel a relatively safe and economical energy source.

1. Typical Residual Fuel Oil Quality Criteria

a. Specific Gravity

This value provides an indication of the potential heating value which can be provided by the residual fuel. The heat energy obtained per unit volume of fuel is greatest for fuels having a high specific gravity.

b. Flash Point

The flash point identifies the minimum temperature at which fuel vapors will ignite. In residual fuel applications, this is helpful in determining whether fuel may be contaminated with high flash materials.

c. Viscosity

This is one of the more important properties of residual fuel oil. It is an indication of both the pumpability characteristics of the fuel and the fuel atomization quality. The viscosity of residual fuel decreases rapidly with increasing temperature. If preheating is available, residual fuels atomize well. If preheating is not available, it may be necessary to burn lower viscosity fuels rather than high viscosity residual oils.

Poor atomization of highly viscous oils can lead to carbonization of the burner tip, carbon in the firebox and overall inefficient combustion. As the viscosity of an oil approaches 1100 cSt, it becomes difficult to pump.

d. Pour Point

The pour point of residual fuel is not the best measure of the low temperature handling properties of the fuel. Viscosity measurements are considered more reliable. Nevertheless, residual fuels are classed as "high pour" and "low pour" fuels. Low pour fuels have a maximum pour point of +60°F (15.5°C). There is no maximum pour point specified for high pour point residual fuels. A residual oil paraffin carbon number analysis is provided in **FIGURE 3-1**.

e. Ash Content

Catalyst fines, metals, rust, sand and other material can be contained in residual fuel. These compounds arise from the crude oil, processing catalysts, water contamination, transportation and storage of the fuel. If the total ash content is >0.20 wt%, deposits can form in burner systems and corrosion in high temperature burners can occur.

Alumina, iron, nickel, silica, sodium and vanadium are examples of compounds which can be found in residual fuel ash. If the vanadium content of residual fuel is high, severe corrosion of turbine blades can occur and exhaust system deposit formation can be enhanced. Vanadium enhanced corrosion can occur at temperatures above 1200°F (648.9°C).

VANADIUM/SODIUM ENHANCED CORROSION

Vanadium/sodium enhanced corrosion of metal occurs when molten slag containing vanadium compounds forms on metal system parts. The steps in the corrosion sequence are listed below:

1) Vanadium and sodium compounds present in the fuel are oxidized during fuel combustion to V_2O_5 and Na_2O.

2) The V_2O_5 and Na_2O adhere to and react with the metal surface and form a molten liquid or eutectic with the metal.

3) The liquid dissolves the magnetite or Fe_3O_4 protective layer on the metal surface. This exposes the underlying metal layer to rapid oxidation, most likely catalyzed by vanadium.

4) The oxidation which occurs in the hot environment rapidly reduces the thickness of the metal components and leads to component failure. Corrosion rates as high as 0.1-0.2 in./year have been demonstrated at 1300°F (704.4°C) operating temperatures.

Compounds of magnesium can be added to fuel containing vanadium to help combat its corrosive effects. Typical addition rates range from 2:1 Mg:V at a 1200°F

(648.9°C) operating temperature and 3:1 Mg/V at a 1500°F (815.5°C) operating temperature. However, the total ash content must still be monitored.

f. Carbon Residue

Carbon number measurement serves as an indication of the tendency of fuel to form deposits in a system where the available air supply is limited. This value is less meaningful for home heating units burning distillate fuel.

g. Water and Sediment

Filter plugging and burner problems can be caused by the presence of water insoluble sediment and waterborne solids in residual fuel. Specifications on water and sediment typically range from 1 vol% to 2 vol%.

h. Sulfur Content

Presently, certain marine residual oils have sulfur specifications ranging as high as 5.0 wt%. Concentrations from 1.0 wt% to 2.0 wt% are more typical. Fuel sulfur can lead to deposits and eventual corrosion problems within transportation, storage and burner systems.

Burning of sulfur to produce SO_x can create both burner system corrosion problems as well as atmospheric air emission concerns. About 1-5% of the fuel sulfur burned is converted to SO_3 and the remainder is converted to SO_2. If a system operates below its dew point, the SO_3 can react with condensed water to form sulfuric acid. Much work is being done through hydrodesulfurization, neutralization and engineering to reduce the amount of sulfur oxides produced through burning of residual fuel.

i. Distillation

This measurement is typically not specified for residual fuels. End points may be over 1000°F (537.8°C).

FIGURE 3-1. Paraffin Carbon Number Analysis of a Typical Atmospheric Tower Residual Oil

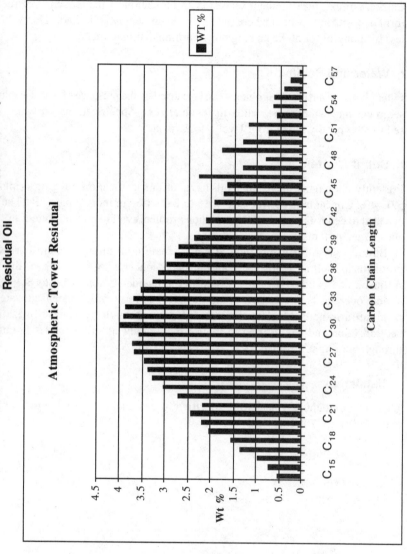

CHAPTER 4

Common Sources of Fuel Performance Problems

A. ENVIRONMENTAL SOURCES

Elements of the environment continually impact fuel performance. The effect of water, cold temperatures, heat, air, light and external contamination can initiate a variety of problems. The first place to begin looking for the cause of a fuel handling or performance problem would be these environmental sources.

1. Water

Water is one of the most abundant, unique and utilized compounds on earth. It functions as a powerful solvent, provides heating and cooling, is a common reagent in most chemical reactions and comprises most of our body weight.

It is unique because of its powerful hydrogen bonding character and its distinct polarity. However, in fuel systems, these characteristics of water make it a source of a variety of problems. Corrosion of metal fuel system components, emulsification with fuel performance additives and ice formation in fuel lines are some of the problems directly related to the presence of water in fuel.

a. Solvent Properties of Water

The ability of water to function as a solvent for many compounds also makes it useful in the fuel refining and processing industry. It is utilized to wash and remove salts from crude oil prior to distillation, remove H_2S and mercaptans from processed fuel (as caustic water) and function in the removal of trace amounts of acid from acid catalyzed fuel processing reactions.

Water can dissolve both organic and inorganic compounds. The presence of these compounds can increase the ability of water to conduct electrical charge. Because of this fact, it is possible to estimate the total dissolved solids (TDS) in water by measuring a change in its ability to conduct electrical charge. The relationship between electrical conductivity and TDS in water is shown in **TABLE 4-1.**

TABLE 4-1. Effect of Total Dissolved Solids on the Electrical Conductivity of Water

Conductivity, pS	TDS, mg/L
1	0.5
25	13.4
50	27.4
100	56.3
500	303
1000	630

The presence of organic compounds in water can occur as natural compounds or as contaminants. Microorganisms and plant decomposition products can occur naturally in water. Alcohols, organophosphates, glycols and organohalides are examples of organic contaminants frequently found in water. Analytical methods including the biological oxygen demand (BOD), chemical oxygen demand (COD), total organic carbon (TOC), total organic halide (TOX) and oil and grease are utilized to measure the quality and concentration of organic compounds in water. These methods are summarized in **TABLE 4-2**.

TABLE 4-2. Methods Utilized to Measure the Concentration of Organic Compounds in Water

Method	Summary
Biological Oxygen Demand (BOD)	A five day incubation is utilized to determine the amount of oxygen consumed when bacteria digest organic matter in water. High BOD may indicate an oxygen shortage in water for fish and other aquatic life
Chemical Oxygen Demand (COD)	Utilizes potassium dichromate to oxidize and measure the amount of oxygen consumed by biodegradable and non-biodegradable organic material and oxidizable inorganic compounds $BOD = (0.7 - 0.8) \times COD$

Total Organic Carbon
(TOC) — The total amount of organic carbon is measured in water by complete oxidation to CO_2 and H_2O. Oxidation methods utilized for measurement include combustion-IR and persulfate-UV

Total Organic Halide
(TOX) — Organohalides are separated from water by adsorption onto a charcoal packed column. The charcoal is pyrolyzed releasing HX. The HX gases are then analyzed by coulometry

Oil & Grease — Mineral oil, lipids, dyes, detergents and elemental sulfur are extracted into hexane or onto a solid phase silica-hydrocarbon medium and separated

b. Water in Fuel Systems

Described below are some of the more common problems cause by the presence of water in fuel systems:

FERROUS METAL CORROSION

Most fuel system storage tanks, transfer lines and underground pipelines are composed of 1018/1020 carbon steel. These system components are all susceptible to internal corrosion whenever fuel containing water is introduced. Other factors which can enhance fuel storage and transportation system corrosion include:

- Acid carryover from fuel processing
- Products of microbial growth
- Sea water/salt water contamination

Whenever deposits from fuel systems are analyzed and are found to contain high levels of iron, corrosion is probably occurring somewhere within the fuel system.

When water pH is <6, iron corrosion and the formation of corrosion products such as colloidal ferric hydroxide can result. Colloidal ferric hydroxide, however, is difficult to remove through filtration and difficult to detect. Fuel containing these particles appears bright and clear. Only about one micron in diameter, colloidal ferric hydroxide compounds can pass through fuel filters and deposit onto fuel system components. Further system corrosion can follow.

ACID CARRYOVER/LOW WATER PH

During the processing of fuels, acids can be used as extracting/neutralizing agents as well as reaction catalysts. For example, sulfuric acid can be used to remove olefins from fuel. Sulfuric/hydrofluoric acids are used as reaction catalysts in the production of high octane gasoline alkylate fractions.

Refiners typically wash acid processed fuel with water or caustic to remove the acid. However, the remote possibility still exists for some acid to carry over with water into downstream operations. If acid remains, the likelihood of fuel system corrosion is markedly enhanced.

CAUSTIC CARRYOVER

Processing of fuel sometimes requires the use of caustic. Neutralization of the acidic alkylation catalysts and the sweetening of sour fuel streams are both accomplished by utilizing caustic.

The carryover of caustic into a finished fuel blend usually has minimal effect alone on the corrosion of ferrous metals. However, in fuel systems containing a conventional tall oil dimer-trimer fatty acid or partially esterified corrosion inhibitor, caustic can react with and negate the effect of the corrosion inhibitor. As a calcium or sodium salt, these inhibitors will no longer function effectively as an oil soluble, fuel corrosion inhibitor.

As salts, these inhibitors can readily interact with water to form emulsions. The ability of the inhibitor to effectively prevent fuel system corrosion is lost. Once emulsified, corrosion inhibitors can initiate other problems such as filter plugging and sticking of moving parts.

$$R\text{-}COOH + NaOH \rightarrow R\text{-}COO^- Na^+ + H_2O$$

REACTION OF AN ORGANIC ACID WITH CAUSTIC (SODIUM HYDROXIDE SOLUTION) TO FORM A WATER SOLUBLE SODIUM SALT OF AN ORGANIC ACID AND WATER

Failure of sensitive filtration tests such as ASTM D-2276, Particulate Contamination in Aviation Fuel by Line Sampling, can be due to caustic neutralized corrosion inhibitor salts. Sodium or calcium salts of dimer-trimer fatty acid corrosion inhibitors are gel-like in character. Filtration of jet fuel containing gelled corrosion inhibitor will be impeded due to plugging of fuel filter media by the inhibitor gel. This slowdown of filtration can result in failure of jet fuel to pass this critical performance test.

WATER INITIATED EMULSIFICATION OF FUEL PROCESSING ADDITIVES

During the refining and processing of fuel, corrosion inhibitors, antifoulants, filmers, neutralizers and other organic compounds may "carry over" into a finished product. These polar organics may attract and interact with water to tightly bind it into the fuel as an emulsion. The result is usually a cloudy, hazy fuel. These emulsions are often quite difficult to break. If the water present contains caustic, organic salts or corrosion products, the emulsion may be quite stable.

These emulsions can eventually settle from fuel and accumulate as a film-like layer on the bottom of fuel storage and distribution systems. The result of this accumulated emulsion can lead to any of the following:

- Metal corrosion
- Site for initiating microbial growth
- Filter plugging/pump sticking
- Particulate contamination (jet fuel)

WATER INITIATED EMULSIFICATION OF FUEL DETERGENTS AND DEPOSIT CONTROL ADDITIVES

Detergents added to gasoline and diesel fuel which help to control the formation and buildup of deposits within automobile and truck engines are highly polar organic compounds such as succinimides and polyethers. Water can have a strong affinity for many of these detergents.

If water emulsifies with these detergents, the emulsion formed is usually heavier than the fuel and falls from solution as a gel-like emulsion. As a consequence, the fuel no longer contains the detergent and will not provide deposit control performance in the fuel.

Also, the gel-like emulsion formed can be pumped with the fuel volume and accumulate on fuel filters. The result of this accumulation is usually filter plugging, pump sticking and possible engine shutdown.

2. Heat

a. Change in Bulk Volume

Petroleum products expand when heated. The coefficient of expansion for most petroleum products will vary and is dependent upon the specific gravity or °API of the product. A listing of expansion coefficients for petroleum products is provided in **TABLE 4-3**.

TABLE 4-3. Expansion Coefficients for Petroleum Products

°API @ 60°F	Expansion Coefficient per °F @ 60°F	Effective °API Range
6.0	0.00035	0 to 14.9
22.0	0.00040	15.0 to 34.9
44.0	0.00050	35.0 to 50.9
58.0	0.00060	51.0 to 63.9
72.0	0.00070	64.0 to 78.9
86.0	0.00080	79.0 to 88.9

An example of how to use the information in this chart is described as follows:

A 6000 gallon load of kerosene at 70°F leaves a fuel terminal by tank truck. Upon arrival at its destination, the fuel has cooled to a temperature of 40°F. The °API of this kerosene @ 60°F is 44.0. What will be the volume of this kerosene at this new temperature of 40°F ?

Calculation: 70 - 40 = 30
30 × 0.00050 = 0.015
6000 × 0.015 = 90
6000 - 90 = 5910 gallons volume @ 40°F

The reduction in kerosene volume from 6000 gallons to 5910 gallons is the result of an actual contraction or shrinking of kerosene as it cools. Likewise, fuel volumes can expand comparatively when heated.

The Expansion Coefficient values per °F listed in **TABLE 4-3** are for the °API ratings shown. Expansion Coefficient values for measurements which fall within the Effective °API Range limits provided in TABLE 4-3 can be extrapolated. A more detailed volume conversion table appears in **APPENDIX 2**.

Expansion and contraction of fuels and oils is common and can explain why product volumes can increase or decrease as temperature changes.

b. Air Temperature - Air Volume Relationships

As air temperature increases, air density decreases. This results in a reduction in the amount of oxygen available per unit volume of air to aid in the combustion of fuel. **FIGURE 4-1** below demonstrates the loss in fuel combustion efficiency as ambient air temperature increases. As shown, a loss of combustion efficiency occurs as intake air temperature increases above 90°F. A similar reduction in efficiency occurs with increasing altitude.

FIGURE 4-1. Reduction in Fuel Combustion Efficiency with Increasing Air Temperature

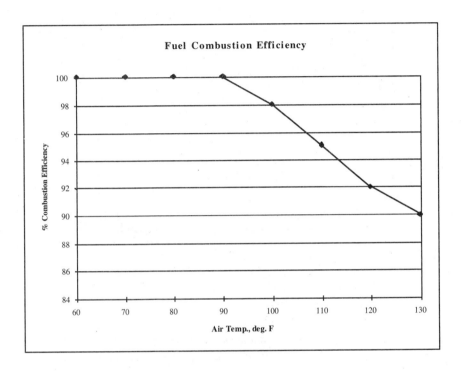

c. Change in Rate of Reaction

Heat increases reaction rates. It is common knowledge that chemical reaction rates double whenever the temperature of a system increases by 10°C or each 18°F. This also means that reaction rates can slow by one-half whenever temperature decreases by the same amount.

Therefore, in environments where high temperatures are maintained for long periods of time, the processes of oxidation and corrosion could be expected to accelerate. The process of demulsification may also be enhanced. **FIGURE 4-2** demonstrates how corrosion of mild steel increases as water temperature increases. Also demonstrated is the reduction in the corrosion rate obtained whenever oxygen is free to escape from an open system.

FIGURE 4-2. Effect of Temperature on the Corrosion of Carbon Steel in Water

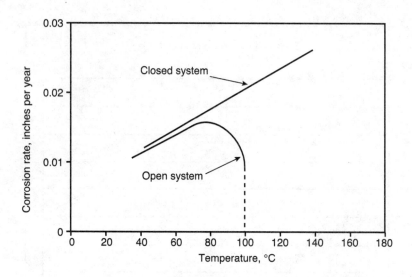

Examples of other problems related to heat are listed below. A detailed description of each of these problems can be found in later chapters of this book.

TYPICAL FUEL PROBLEMS ENHANCED BY OR RELATED TO HEAT

- Evaporation and loss of fuel components upon storage; resultant loss of fuel performance
- Increase in gum formation of stored gasoline and diesel fuel
- Degradation of diesel fuel color and accumulation of insoluble deposits
- Increase in the rate of metal corrosion
- Possibility for enhancement of microbial growth 90°F (32.2°C) to 100°F (37.8°C)

3. Air and Oxidation

Air contains approximately 21% oxygen and 78% nitrogen and <1% of other elements such as carbon dioxide and argon. Since oxygen is known to react with numerous materials, it can be a source of fuel problems. Hydrocarbons which have reacted with oxygen from air can yield a variety of compounds such as peroxides, organic acids and gums.

a. Deposit Formation

Most of the deposits formed in fuel and oil systems are rich in oxygen content. Oxygen in these deposits is not naturally occurring in the petroleum product, but typically comes from atmospheric oxygen.

When oxygen reacts with fuel components, the fuel first begins to darken in color. This is especially true for fuel components possessing conjugated double bonds. The oxidized hydrocarbons which result tend to further react to form polymeric compounds. These polymeric compounds are high in molecular weight, usually fuel insoluble and possess a slight negative charge. The tendency of these compounds to form sludge and varnish-like deposits is great. These deposits can "stick" tightly to metal fuel system parts such as bulk storage tank surfaces, pipeline walls and transfer lines.

The following general rule applies to the oxidation of petroleum hydrocarbons:

- Paraffins oxidize to form acidic compounds
- Naphthenic and aromatic compounds oxidize to form high molecular weight deposits
- Olefins participate in a variety of oxidation reactions which lead to the formation of carbonyl compounds, gums and high molecular weight deposits

Additional information related to fuel oxidation and deposit formation appears later in this chapter.

4. Cold Temperature Conditions

Low temperatures can be the source of a wealth of fuel problems. When fuels and oils are cold, they can:

- Gel
- Contract
- Become cloudy or hazy
- Increase significantly in viscosity
- Become exposed to condensed water
- Precipitate wax
- Cause pumping problems due to failure to flow
- Become less volatile

All of these phenomena can result in operating problems with machines, engines, pumps and other equipment utilizing petroleum products.

a. Gellation

Some petroleum products, especially those containing higher molecular weight compounds such as waxes, do not crystallize rapidly when cooled. Instead, they form a gel-like network throughout the fuel matrix. This network can begin forming at temperatures well above the pour point of a fuel and may render the product unpumpable.

It is recommended that any higher viscosity product such as residual oil or heavy distillate fuel be evaluated for changes in low temperature handling properties over time. Testing for reversion in pour point by the Shell Amsterdam Reversion test or the British Admiralty Pour Point Reversion test are recommended. Also, viscosity increase vs. temperature decrease determinations are recommended for products stored at low temperatures for extended periods of time.

b. Contract

As previously discussed, the coefficient of expansion/contraction for petroleum products is dependent upon specific gravity. As fuels cool, their volume will be reduced.

Unlike ice, solidified fuels do not expand to form well organized and structured crystals. Therefore, gelled or solidified fuel, such as diesel fuel and residual fuel oil, will not expand within the container in which it is stored. It will contract and collapse slightly, pulling away from the side walls of the tank or container in which it is held.

c. Become Cloudy or Hazy

Low levels of water, up to approximately 50-75 ppm, can be solubilized into most middle distillate fuel at ambient temperature without causing the fuel to appear hazy. However, as fuels cool, dissolved water becomes visible as haze. Conversely, as fuels are heated, water haze disappears.

The data in **FIGURE 4-3** demonstrates the effect of temperature on the solubility of water in distillate fuel containing increasing concentrations of water.

FIGURE 4-3. Water Solubility in a Typical Low Sulfur Diesel Fuel

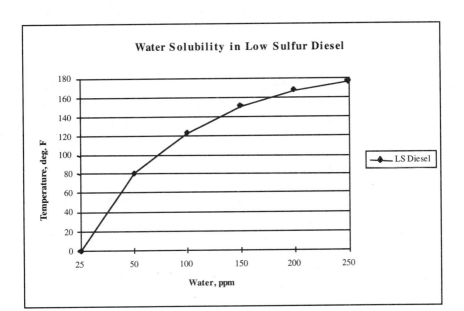

As an example, from the information in FIGURE 4-3, it can be determined that a distillate fuel at 75°F can solubilize 50 ppm of water without appearing cloudy or hazy. At the same temperature, fuel containing 100 ppm of dispersed or emulsified water may appear hazy.

d. Increase Significantly in Viscosity

As temperature decreases, fuel viscosity increases. This fact can lead to a variety of problems related to the pumping, mixing and utilization of fuels in cold environments. Pump cavitation and starvation can occur if cold, viscous fuel does not flow to the pump. Vane pumps, gear pumps and piston pumps all rely on the material being pumped to serve as both a pump lubricant and as an agent for removing frictional heat. If fuel does not reach a pump due to its high viscosity, pumps can be severely damaged.

e. Become Exposed to Condensed Water

As water vapor in the air cools and reaches its dew point, it begins to condense and fall from the air. Condensation of water vapor on the inside of fuel storage and transportation equipment can lead to unexpected water related problems such as emulsification and bacterial growth. Fuel system condensate water accumulation is a slow process, but over time a significant amount of water can accumulate. Water enhanced problems such as rusting, microbial growth and emulsification can follow.

CARBURETOR ICING

This phenomenon is due to water vapor condensing and freezing on carburetor parts. It occurs when ambient temperature is slightly above freezing and humidity is high. As fuel pumped into the carburetor vaporizes, heat is removed from carburetor parts by evaporating fuel. The carburetor then cools to the extent that water vapor in the intake air freezes as ice around the venturi and other carburetor parts. As ice builds, the flow of air and fuel through the carburetor is restricted and the engine stops.

The icing problem can often "go away" when the engine stops. This is due to melting of ice by the radiant heat emitted by warmer engine parts.

f. Precipitate Wax - Fuel Flow Restriction

Fuels such as diesel fuel and heating oil are sometimes stored in large tanks for extended periods of time. At temperatures below the cloud point of the fuel, wax can form and fall from solution. Accumulated wax within fuel systems can deposit onto component parts and settle into areas of low turbulence. Problems such as filter plugging and flow limitations can be due to accumulated wax.

g. Become Less Volatile

Fuel volatility is an extremely important factor related to fuel combustion and burning efficiency. Evaporation, vaporization and vapor pressure of fuel can all be reduced in cold environments. Poor startability and warmup of gasoline and diesel engines can be directly related to fuel volatility. Also, cold kerosene will not vaporize and burn as efficiently in wick-fed systems.

WHITE SMOKE

Upon starting a cold engine in cold weather, it is not unusual for exhaust to appear as white smoke. This is because the combustion chamber and exhaust system temperatures are not hot enough to enable water and uncombusted fuel to exhaust as hot vapor. They, instead, appear as a cooler, condensed white smoke. After the engine reaches operating temperature, the white smoke usually disappears.

LUBRICATING OIL DILUTION BY FUEL

Higher boiling, less volatile components in gasoline may not completely vaporize during cold weather startup of an engine. In the cylinder, the still liquid fuel components may flow past the piston rings and into the lubricating oil. This process can lead to removal of some of the lubricant from the piston and cylinder walls and also lead to a reduction in the lubricating oil viscosity. Together, these two factors can interfere with the ability of the lubricant to prevent engine wear in cold weather startup.

5. Light and Its Effect

Fuel performance problems initiated by light are not common. However, fuel quality can be affected. The primary concern of light exposure is fuel color darkening and the possible formation of high molecular weight deposits due to free radical initiated polymerization of fuel components.

Light activated shift of double bonds in organic compounds can promote the reaction of oxygen with a molecule. Once oxygen has bonded to a fuel component, oxidized organic compounds such as alcohols, aldehydes, esters, ethers and acids can form. These compounds can then continue to react with other fuel components to form color bodies, gums and insoluble deposits.

Light is an excellent catalyst for the promotion of free radical reactions. For example, the light initiated reaction of an olefin with a mercaptan to form an organosulfide can occur in fuel as follows:

Light Catalyzed Reaction

Phenols, cresols and quinolines are examples of compounds found in fuel which can degrade in color in the presence of light and air. If present in fuel, these compounds can readily degrade the ASTM color or Saybolt color rating of the fuel. These compounds are shown in **TABLE 4-4.**

TABLE 4-4. Effect of Light on Fuel Components

Alkyl Phenol $R > C_1$	Phenols turn pink or red if not perfectly pure or when exposed to light. This change is hastened in an alkaline environment.
Cresols $R_{1,2,3}$ = Methyl R_1 =Ortho; R_2 =Meta; R_3 =Para	Cresols become dark, yellow-brown or red in color on exposure to light and air. *O-Cresol-* colorless or yellowish liquid *M-Cresol-* crystals or liquid becoming dark with age and exposure to air and light.
Alkyl Quinoline	Quinolines are hygroscopic and darken in color when exposed to light.

6. Rust, Dirt and Debris

Fuels are handled in both closed and open systems. Once crude oil reaches the refinery, it is typically held within an entirely closed system. Outside of the refinery, fuels can be openly exposed to the environment. Upon loading into tankers, barges and storage containers, fuels may be exposed to the open environment. Also, after delivery and sale, fuels are often stored in tanks, containers and cans for various lengths of time. During storage, exposure of the fuel to environmental contaminants is quite possible.

Within the closed system of a refinery, rust, metal salts and catalyst fines exist. While being refined, processed and blended, fuels may "pick up" some of these materials and carry them throughout the fuel distribution system. At some later time, these materials can lead to problems such as fuel darkening, combustion system deposits or filter plugging.

In open systems such as drumming operations, barge hold loading and above and underground tank storage, fuels can be exposed to a host of potential contaminants. Rust, dirt, sand, cleaning compounds, metal fines, elastomer degradation products, etc. are some of the contaminants which can be found. These materials can result in fuel problems such as pump wear, emulsion formation, color degradation, combustion system deposits and filter plugging.

Although difficult to predict, the problems related to rust, dirt and debris are usually the easiest to identify.

B. WAX IN PETROLEUM PRODUCTS

Wax related problems are common throughout the petroleum product industry. Fuels and lubricants contain wax at varying concentrations. Filter plugging, line blockage, viscosity increase and product haziness are all symptoms of wax formation within a fuel or oil.

A *wax* can be defined as a linear, branched or cyclic hydrocarbon typically containing from 17 to 60 carbon atoms. Low carbon number waxes are found in middle distillate fuels and typically constitute a low percentage of the paraffins found in distillate fuel. Higher carbon number waxes can be found in residual fuels and lubricating oil. The percentage of wax in residual fuel oils can vary widely depending upon the refining processes utilized.

Refining processes which can influence the wax related character of fuels and oils include the following:

- Composition and character of the crude oil(s) refined
- 90% distillation temperature
- Hydrotreating and hydrocracking aromatics and asphaltenes to paraffins

1. Classifications of Wax

Three classifications of wax as shown in **Figure 4.4** have been defined and include the following:

Plate - This wax structure is the most common and consists primarily of linear *n*-paraffin molecules. The linearity of the wax molecules enables the wax to accumulate to form a uniform crystal lattice. The wax molecules actually "stack" together into a geometric form resembling a flat plate.

Mal-shaped - These wax formations are not as symmetrical as plate crystals. Mal-shaped wax is composed of both linear and branched paraffin molecules. The branched paraffin molecules change the confirmation of the wax so that "stacking" and alignment typical in plate wax formation does not occur. The result is a wax crystal which is smaller than plate crystal wax. The temperature at which mal-shaped wax forms is also lower than plate crystal formation temperature.

Needle - The shape of these wax crystals resembles a hollow cylinder or cone. Needle wax crystals form at the lowest temperature of all wax crystals and are quite small. Cycloparaffins and branched paraffins comprise the needle wax crystal structure.

FIGURE 4-4. Wax Crystal Classifications

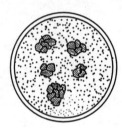

Plate Crystals   *Mal-shaped Crystals*   *Needle Crystals*

MACROCRYSTALLINE AND MICROCRYSTALLINE WAX

Macrocrystalline wax is larger in size and typically contains waxes from 17 to approximately 30 carbons in length. Plate structure wax is macrocrystalline in form and can be found in low concentration in some distillate fuel. Heavy distillates and residual oils can contain higher concentrations of macrocrystalline wax.

The term *microcrystalline wax* is commonly used to describe wax which is either mal-shaped or needle-like. This wax typically contains molecules >30 carbons in length. This wax can be present in various high boiling fractions such as residual fuel oil and lubricant fractions.

Within given molecular weight limits, wax components having higher melting points crystallize in plates. Low melting components crystallize in needles. Others crystallize in mal-shaped conformation. This process is independent of the fuel or oil in which the crystals form.

2. Diesel Fuel Wax Related Problems

At low temperatures, the wax in most distillate fuels can cause severe problems with fuel pumping and filtration. As diesel fuel cools, wax begins to crystallize and become visible. The temperature at which crystallized wax becomes visible is known as the *cloud point*.

This wax can accumulate on fuel filter media and can lead to plugging of small orifices and lines. This plugging temperature can be measured and is commonly referred to as the *filter plugging temperature*. Testing methods utilized to predict the filter plugging temperature and the low temperature flow properties of distillate fuel are listed in **TABLE 4-5**.

TABLE 4-5. Common Test Methods Utilized to Predict the Plugging Tendency of Distillate Fuel

Test Method	Designation
Enjay Fluidity	Industry Standard
Cold Filter Plugging Point (CFPP)	IP 309
Low Temperature Flow Test (LTFT)	ASTM D-4539
Simulated Filter Plugging Point (SFPP)	Proposed European Standard

Further cooling of the fuel leads to wax crystal formation throughout the fuel matrix. The growing wax crystals develop into a larger lattice-like network encompassing the bulk fuel volume. This lattice-like network eventually causes the fuel to become highly viscous and to eventually gel into a semi-solid mass. The lowest temperature at which fuel remains in the liquid state just prior to gellation is called the *pour point*.

a. Effect of Blending Diesel Fuel with Kerosene and Other Light Distillates

CLOUD POINT REDUCTION

It is possible to dilute diesel fuel such as No. 2-D low sulfur with kerosene, #1 fuel oil or jet fuel to reduce the fuel cloud point. Also, additives are also marketed which have the ability to inhibit nucleation of wax crystals in some fuels, thereby lowering the cloud point of the fuel. These products are called *cloud point improvers*.

The use of kerosene or #1 fuel oil in reducing the cloud point of distillate fuel is common practice. By diluting the fuel with these lighter streams, wax related problems can be minimized, but not eliminated. As a general rule a reduction in diesel fuel cloud point by 2°F to 4°F (about 1°C to 2°C) is typical for each 10% of kerosene added. This is illustrated in **FIGURE 4-5**.

FIGURE 4-5. Effect of Kerosene Blending on the Cloud Point of Low Sulfur Diesel Fuel

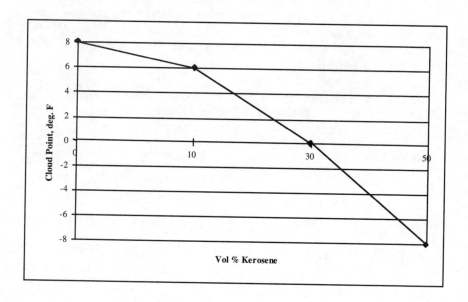

During winter and under other low temperature operating conditions, fuel cannot be effectively filtered at temperatures much below its cloud point unless the fuel wax is diluted with kerosene or treated with a wax crystal modifier.

POUR POINT REDUCTION

Kerosene can be utilized effectively to reduce the pour point of most distillate fuels. The dilution limits are often based upon whether kerosene dilution will negatively impact fuel properties such as the viscosity, distillation parameters, sulfur limit or cetane number.

As a general rule the pour point of a diesel fuel can usually be reduced by 5°F to 10°F (about 3°C to 5°C) for each 10% of kerosene added. The typical maximum blending volume of kerosene is about 30% by volume.

The effect of kerosene dilution on the pour point of a typical low sulfur # 2 diesel fuel is demonstrated in **FIGURE 4-6**.

Common Sources of Fuel Performance Problems 83

FIGURE 4-6. Effect of Kerosene Blending on the Pour Point of Low Sulfur Diesel Fuel

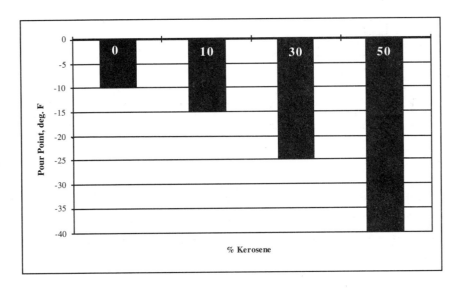

COLD FILTER PLUGGING POINT (CFPP) REDUCTION

The cold filter plugging point or CFPP of a typical #2 diesel fuel can be reduced by the addition of kerosene. The maximum blending volume is again limited by the effect kerosene will have on specific physical and performance properties of the #2 diesel fuel. As a general rule diesel fuel CFPP can be reduced by 2°F to 4°F (about 1°C to 2°C) for each 10% of kerosene added.

The effect of kerosene dilution on the CFPP of a typical #2 diesel fuel is shown in **FIGURE 4-7.**

FIGURE 4-7. Effect of Kerosene Blending on the CFPP of Low Sulfur Diesel

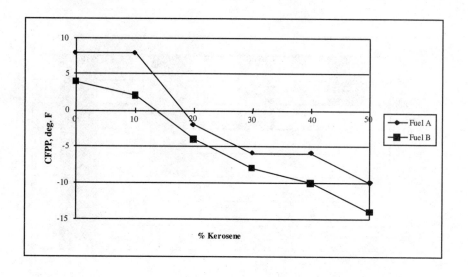

LOW TEMPERATURE FLOW TEST (LTFT) REDUCTION

Reduction in the LTFT is possible with the addition of either kerosene or additives termed low temperature operability improvers. Prediction of response is not readily estimated and testing is always recommended. Typically, kerosene addition will consistently provide a reduction in the LTFT temperature. Operability additives can often be used as a substitute for kerosene or in combination with kerosene to improve the LTFT performance of No. 2-D fuel.

Operability additives must often be used at high treat rates, 1000 ppm or higher, to obtain a reduction in LTFT temperature greater than 10°F (5.6°C).

Some wax crystal modifiers provide LTFT performance. However, as with operability additives, the performance should always be evaluated before use.

b. Wax Related Filtration Problems

Diesel fuel pumped from storage tanks, through pipelines and within internal combustion engines passes through various types of fuel filters. When cold fuel containing wax passes through these filters, the wax can be trapped on the filter media. Accumulated wax can slow the flow of fuel enough to shut down further fuel flow.

Different steps are taken to prevent and avoid this situation. Utilization of low cloud point diesel fuel and blending with kerosene are two possible solutions. Sometimes these measures are viewed as costly and impractical.

The addition of a wax crystal modifier to diesel fuel is a common and a well accepted alternative to kerosene dilution. Wax crystal modifiers are typically polymeric compounds which have the ability to co-crystallize with wax to alter the size, shape and structure of the wax crystal lattice.

Two possibilities exist to explain how wax crystal modifiers work. They are summarized as follows:

SAME TEMPERATURE CRYSTALLIZATION - This concept deals with the possibility of the wax crystal modifier acting as the seed crystal onto which fuel wax crystallizes. This explanation requires that the modifier crystallize at the same temperature as the fuel wax. Once crystallized, the wax crystal modifier thus controls the conformation and structure of the resultant wax crystal.

LATTICE CO-CRYSTALLIZATION - As wax crystals begin to develop, they form an interlocking lattice-like network throughout the fuel. By co-crystallizing with a forming wax lattice, modifiers prevent wax crystals from growing. The conformation and functional groups of the wax crystal modifier prevent the adherence of new wax molecules onto a growing lattice. As a result, smaller wax crystals form and fuel will continue to flow at very low temperatures.

c. Distillate Fuel Wax Crystal Modifiers

The most common type of wax crystal modifier used to reduce the pour point and filtration temperature of distillate fuel is based on ethylene vinylacetate (EVA) copolymer chemistry. These compounds are quite common throughout the fuel additive industry. The differences, however, are found in the variation in the molecular weight and the acetate ratio of the copolymer.

On occasion, the performance of an EVA copolymer can be enhanced by blending with a wax crystal modifier of a different chemical type. Wax crystal modifiers used to modify the crystal structure of lubricant, residual fuel and crude oil waxes can be blended at low concentrations with EVA copolymers to improve their performance. However, the performance enhancement is usually fuel specific and not broad ranged. Also, the low temperature handling properties of the EVA may be impaired when blended with other wax crystal modifiers.

Problems associated with the use of wax crystal modifiers do not pertain so much to the ability of the modifier to perform, but to the proper application technique. These copolymers are quite viscous in nature and must be diluted in solvent in order to be handleable. Even after dilution, they are still quite viscous and have relatively high pour points.

Whenever a wax crystal modifier does not perform as expected, there are several possible explanations. Outlined below are some of the possibilities:

FUEL WAS BELOW ITS CLOUD POINT

In order for a wax crystal modifier to function properly, it must be present to co-crystallize with fuel wax. This requires that the modifier must be added to fuel well before wax crystal formation begins.

It is known that wax can begin the process of organization into a crystal structure above the actual, observable cloud point temperature. Because of this fact, the wax crystal modifier should be added at a temperature at least 20°F (11.1°C) *above* the cloud point of the fuel. Addition at this higher temperature helps to ensure that the modifier is completely solubilized in the fuel prior to the formation of the wax crystals.

Wax crystal modifiers added *after* wax crystals begin to form will have only minimal affect at modifying the size, shape and structure of wax crystals. Consequently, little improvement in the low temperature handling characteristics of the fuel will be obtained.

ADDITIVE WOULD NOT DISSOLVE IN COLD FUEL

When wax crystal modifiers are added to cold fuel, even to fuel well above its cloud point, modifiers may not dissolve properly. The polymeric nature of wax crystal modifiers makes them quite viscous at low temperatures. Additive suppliers will often provide modifiers in a highly dilute form (i.e., 10% or 20% solution) so they will remain fluid at low temperatures.

However, if a typical, non-diluted wax crystal modifier is added to fuel which is at a cold temperature of +10°F (-12.2°C) to +20°F (-6.7°C), it may not dissolve completely in this fuel. The result will be additive accumulation as a viscous layer at the bottom of a fuel or storage tank. Ultimately, the additive will be trapped by a filter as it flows from the tank.

A second, and even worse possibility, would be the addition of cold additive to cold fuel. In this case, the additive would not dissolve at all and would set as a gelled mass at the bottom of the fuel tank. When the gelled wax crystal modifier does move from the tank, it may plug a cold fuel line or filter. If allowed to reach a pump, the gelled additive could cause sticking of pistons or other pump parts.

FUEL WAS PREVIOUSLY TREATED WITH A WAX CRYSTAL MODIFIER

Occasionally, wax crystal modifiers will not provide the performance anticipated. Either the response to additive treatment was much less than expected or no response was obtained.

When this occurs, it is quite possible that the fuel was previously treated with a wax crystal modifier. Under these circumstances, the expected performance of secondary treatment with wax crystal modifier is minimal.

It is quite time consuming and expensive to analyze for the presence of a wax crystal modifier in fuel. However, it is possible to determine whether a fuel already contains a wax crystal modifier by analyzing the following test information:

Compare cloud and pour point values - Fuel which does not contain a wax crystal modifier will have temperature differences between the cloud and pour points typically from 15°F (8.3°C) to 20°F (11.1°C). If the difference between the cloud and pour point values is greater than 25°F (13.9°C), it is quite reasonable to believe that the fuel contains a wax crystal modifier.

Compare cloud and CFPP values - The cloud point and the cold filter plugging point temperatures for fuel which does not contain a wax crystal modifier can often be the same. Typically, untreated cloud point and CFPP values will be within 2°F to 4°F (about 1°C to 2°C) of each other. If the temperature difference between an untreated fuel's cloud point and CFPP differ by 10°F (5.6°C) or more, the fuel probably contains a wax crystal modifier.

3. Problems With Reproducing the Pour Point of Crude Oil and Heavy Residual Products

Testing of the pour point of crude oil and certain residual fuel products requires an understanding of how these oils can behave under certain conditions of heating and shearing.

a. Effect of Applied Shear

When pour point testing is performed on crude oil, little to no shear is applied to the oil in the pour point tube. Under these conditions, the wax crystal lattice matrix which forms in the crude oil normally remains intact and the oil gels at the pour point.

However, when disturbed by pumping, mixing or agitation, the loosely formed wax crystal lattice can sometimes be broken with applied shear. If this occurs, some crude oils may again begin to pour and continue to flow at temperatures below the initial reported pour point.

b. Effect of Heating

CRUDE OIL

If shearing has destroyed the loosely formed wax lattice network of gelled crude oil so that the oil flows below its natural pour point, heating can restore the oil to its original pour point. By heating the crude oil to temperatures 20°F (11.1°C) to 30°F (16.7°C) above the cloud point, waxes can be melted, solubilized and redistributed into the oil. When the pour point is then determined for this heated oil, the result obtained may be higher than the result obtained for the same oil which was not heated prior to pour point testing. All wax must be melted and solubilized into crude oil prior to pour point testing. Otherwise, erroneous and often lower crude oil pour point values may be obtained.

RESIDUAL OIL

When determining the pour point of certain heavy residual products such as #6 fuel oils, bunker fuels, vacuum gas oils, vacuum resids, atmospheric resids and visbreaker bottoms, it is important to pay close attention to the temperature applied to the oil prior to pour point testing. In some cases, preheating an oil to temperatures greater than 212°F (100.0°C) prior to pour point testing can result in a pour point value which is lower than the value obtained for the same oil preheated to 110°F (43.3°C).

It is believed that asphaltic compounds within these heavy oils may be better distributed by high temperature preheating. Distribution of these compounds may interfere with the formation of an organized wax lattice throughout the oil matrix as the oil cools. The result would be a pour point which is lower than that for an unheated or mildly heated oil.

C. LOW VOLATILITY

Volatility is important to consider when discussing the combustibility of gasoline and the burning quality of kerosene. Gasoline volatility is crucial to the combustion process. In order to initiate and ensure smooth combustion of gasoline, volatile compounds such as low molecular weight branched paraffins and aromatics must be present. Factors important to consider concerning gasoline combustion are provided as follows:

1. Reid Vapor Pressure

Reid vapor pressure is measured at 100°F and is used to help ensure that gasoline will vaporize adequately and ignite within the combustion chamber of an engine. Vapor pressure is provided by volatile gasoline components such as dissolved butane gas and the presence of pentanes, hexanes, heptanes and benzene.

Because these compounds have a relatively high vapor pressure, they can escape naturally from gasoline. When gasoline is stored for time periods longer than six months, it can become ineffective because it has lost these lighter, high vapor pressure compounds. Consequently, engine startability and warmup are poor. The vapor pressure values of various light hydrocarbons at different temperatures are provided in **FIGURE 4-8**.

FIGURE 4-8. Vapor Pressure of Volatile Hydrocarbons Found in Gasoline

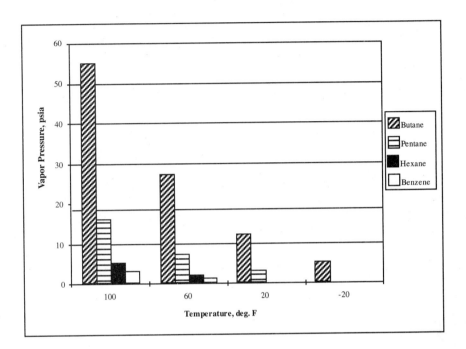

2. 50% Distillation Point

This temperature can be used to help identify whether gasoline will perform well during the warmup period. If the 50% point is high, approaching 300°F (148.9°C) for example, engine warmup performance will be poor. Aged or stored gasoline may lose enough volatile components so that startability and warmup time may be seriously impacted.

3. Percent Distilled @ 158°F

Problems can occur if volatile gasoline components are lost due to evaporation or if gasoline is contaminated with any compound which affects a change in the volatility characteristics of the fuel. For example, engine warmup time will be extended if the relationship between the outside temperature and the percent gasoline distilled at 158°F (70°C) is exceeded.

By using the information in **TABLE 4-6,** it can be determined that engine warmup time problems may result if the percent distilled @ 158°F volume is *less than* the reported volumes at the given temperatures.

TABLE 4-6. Ambient Temperature and Gasoline Distillation Parameters Affecting Engine Warmup Time

Percent Distilled @ 158°F	Minimum Outside Temperature, °F
3	80
11	60
19	40
28	20
38	0
53	-20

Refiners do change the distillation parameters of their gasoline to meet expected regional changes in temperature. However, these changes are made on seasonal averages and are not typically altered to meet a sudden drop in temperature. Using this information, it is easy to explain why rough running and warmup problems are noted during the fall months when early season cold weather changes cause temperatures to suddenly drop.

4. Initial Boiling Point (IBP)

Kerosene used for illumination purposes must contain enough light end, volatile compounds to freely travel up wick-fed burners. If the initial boiling point and 10% point of kerosene are well above 350°F (176.7°C), the fuel will probably have poorer illuminating properties than a fuel with a 310°F (154.4°C) initial boiling point. Also, kerosene which contains higher boiling components can carbonize and degrade lamp wicks. A combination of proper volatility and viscosity help ensure the proper burning characteristics of kerosene in wick-fed systems.

D. DIESEL FUEL CETANE NUMBER AND CETANE INDEX DETERMINATIONS

Although cetane engine testing has been used for many years to evaluate diesel fuels for their cetane number, confusion still exists with regard to this evaluation method. Problems associated with cetane number testing are usually attributed to the following:

- Confusion between cetane index and cetane number
- Unrealistic expectations from a cetane improver
- Reproducibility differences between operators testing the same fuel
- Effect of kerosene blending

1. Cetane Index and Cetane Number

The calculated cetane index methods were developed to help predict the fuel cetane number without extensive engine testing. Since the calculated cetane index is determined from physical property values such as the specific gravity and the 10%, 50% and 90% distillation points, it is readily measurable from laboratory data. The ASTM methods D-976 and D-4737 are used to determine the diesel fuel cetane index.

The cetane number, however, is an engine measurement that requires a skilled operator and a well performing engine to determine. Because of the variability among engines and the differences in operator skill, engine number determinations can differ.

As a general rule, it is usually safe to assume that the calculated cetane index will always be higher than the actual cetane number. On rare occasions, the numbers will be the same, but usually not. Because of this, if a calculated cetane index value is known, it is reasonable to believe that the cetane engine number is usually *one to two numbers lower* than the calculated cetane index.

For this reason, refiners often set cetane index specifications higher than the desired number by one to two numbers or more.

2. Realistic Performance Provided by Cetane Improver

Throughout the world, the most commonly utilized and available cetane improver is based on isooctylnitrate chemistry. Although other compounds have been shown to provide better performance, their cost, availability and potential safety hazards make them undesirable.

Isooctylnitrate cetane improvers effectively boost the cetane number by aiding in the autoignition of fuel, thus controlling ignition delay. Some typical guidelines which can be used to anticipate cetane improver performance are listed below:

- Low base cetane fuel will not respond to cetane improver treatment as effectively as higher base cetane number fuel. This is demonstrated in **FIGURE 4-9.**

FIGURE 4-9. Cetane Improver Response in a Base 32 Cetane and a Base 42 Cetane Fuel

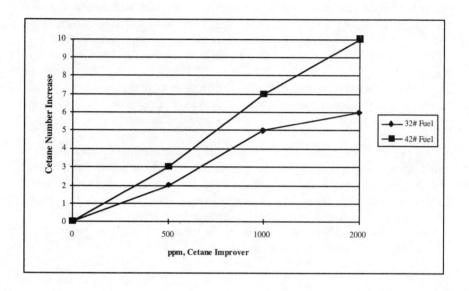

- The greatest boost in cetane number improvement is obtained within the first 1000 ppm of additive treatment. This is also demonstrated in FIGURE 4-9.

- A one to two number cetane number increase is typically achieved for every 250 ppm of cetane improver added to diesel fuel. On rare occasions, a three number boost is obtained for each 250 ppm, but this should not be expected.

- In most fuels, a maximum boost of no more than twelve to fourteen cetane numbers from the base number should be expected. In exceptional cases, a boost greater than fourteen cetane numbers can be obtained. This extreme level of performance improvement, however, is not common.

- Cetane improver addition will not change the calculated cetane index, only the engine cetane number will change.

- In Federal diesel fuel oil DF-A, cetane improver concentration cannot exceed 0.25 wt% and in DF-1 and DF-2, cetane improver concentration cannot exceed 0.5 wt%.

3. Reproducibility Differences Between Different Operators Testing the Same Fuel

Diesel cetane number reproducibility concerns usually develop whenever motor cetane numbers differ by more than one number. The reproducibility variation arises due to operator and engine differences. The typical reproducibility variance accepted by ASTM will change with increasing cetane number. The following reproducibility and repeatability limits have been established by ASTM for method D-613.

Average Cetane Number Level	Reproducibility Limits, Cetane Number	Repeatability Limits, Cetane Number
40	2.5	0.6
44	2.6	0.7
48	2.9	0.7
52	3.1	0.8
56	3.3	0.9

Cetane engine number determinations between operators in different laboratories do vary. If the variance is within the limits shown above, the variance is acceptable and should not be regarded as erroneous. If the difference is outside of the reported limits, then the results can be fairly questioned.

Repeatability limits pertain to the difference between two test results obtained by the same operator for the same test fuel under constant operating conditions. If test values fall outside of these limits, the repeatability is considered poor.

4. Effect of Kerosene Blending

Unfortunately, cetane engine number values for kerosene vary from below 40 to about 50. For this reason, cetane engine testing is always recommended when kerosene is blended into diesel fuel. Hydrotreated kerosene will probably have a higher cetane number than more aromatic kerosene.

5. Ignition Delay and Cetane Number

Since diesel fuel combustion is an autoignition process, the quality and characteristics of the fuel can have a dramatic impact on the efficiency of engine operation. Ignition delay and cetane number go hand-in-hand to influence the quality of the fuel combustion process. They are the primary engine operability parameters directly related to fuel composition.

Ignition delay can be defined as the time period between the injection of fuel into the combustion chamber and ignition of the fuel. A long ignition delay period is characteristic of low cetane fuels; a shorter period is a characteristic of higher cetane fuels.

Ignition delay between injection of fuel and autoignition can be controlled by engine design, fuel and air temperature, fuel atomization and fuel composition. The ignition delay period is typically shorter than the fuel injection period. The effect of temperature on ignition delay is shown in **Figure 4-10**.

FIGURE 4-10. Effect of Temperature on Ignition Delay

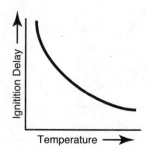

Fuel cetane number is one of the basic characteristics required to ensure proper engine operation and influence ignition delay. The following fuel properties affect cetane number and ignition delay:

- High cetane fuels burn with a shorter ignition delay and lower peak pressure.
- High cetane fuels typically burn with less smoke and odor.
- Cold temperature starting is improved with high cetane fuels.
- Straight chain paraffins autoignite more readily than branched chain paraffins and aromatics of the same carbon number.
- Engine knock and vibration can be due to the rapid pressure rise resulting from the combustion of low cetane number fuels.

6. Smoke and Hydrocarbon Emissions from High Cetane Fuel

This atypical problem results when fuel injected into the combustion chamber ignites and burns before properly mixing with air. Injection of fuel too early in the compression cycle is more of a mechanical problem than a fuel performance problem. It can occur when engines timed to run effectively on low cetane fuel are switched to operate on high cetane fuel.

If fuel autoignition begins before desirable combustion air pressures and temperatures are reached, less fuel carbon is oxidized during combustion and fuel efficiency is lost. For this reason, the fuel injection timing must always be monitored and adjusted to ensure proper mixing of fuel with fully compressed and heated air. Without retarding the injection timing to account for the shorter ignition delay character of high cetane fuels, exhaust smoke and hydrocarbon emissions may result.

7. Cetane Engine Reference Fuels

Either primary or secondary reference fuels can be utilized in the cetane engine when determining the cetane number of distillate fuel. Primary reference fuels are *n-cetane* and *heptamethylnonane*. Secondary reference fuels are identified as *T Fuel* and *U Fuel*. The characteristics of each of these fuels are outlined in TABLE 4-7.

TABLE 4-7. Characteristics of Primary and Secondary Cetane Reference Fuels

Reference Fuel	Characteristics
n-Cetane	Cetane number of 100
Heptamethylnonane	Cetane number of 15
T Fuel	Typical cetane number between 73 & 75
U Fuel	Typical cetane number between 20 & 22

The T Fuel and U Fuel references are commonly used for routine cetane engine determinations. Care should be taken to ensure that both T and U reference fuels are not exposed to long-term storage at temperatures below 0°C (32°F). Both fuels may gel and solidify at low temperatures.

E. HYDROPEROXIDES AND OLEFINS

1. Hydroperoxides

Certain fuel components can be oxidized by the free radical process. **Benzyl, allyl and tertiary** compounds can all be oxidized at room temperature to yield hydroperoxide species.

Through the process of *hydrogen abstraction*, and consequential formation of peroxyradical and alkyl radicals, fuel components can be completely consumed through oxidation. The chain reaction which progresses through the formation of organic hydroperoxides can yield insoluble gums and sludge.

a. Hydrogen Abstraction

$$-\overset{\backslash}{\underset{/}{C}}:H \;+\; O=O \;\rightarrow\; -\overset{\backslash}{\underset{/}{C}}\cdot \;+\; HOO\cdot$$

Tertiary   Oxygen   Alkyl   Hydroperoxide
Hydrocarbon   Radical   Radical

The newly generated alkyl radical can react with oxygen to produce a peroxy radical:

$$-\overset{\backslash}{\underset{/}{C}}\cdot \;+\; O=O \;\rightarrow\; -\overset{\backslash}{\underset{/}{C}}\text{-O-O}\cdot$$

Peroxy radical

The peroxy radical can further react with a paraffin to form an alkyl hydroperoxide:

$$-\overset{\backslash}{\underset{/}{C}}\text{-O-O}\cdot \;+\; -\overset{\backslash}{\underset{/}{C}}:H \;\rightarrow\; -\overset{\backslash}{\underset{/}{C}}\text{-O-O-H} \;+\; -\overset{\backslash}{\underset{/}{C}}\cdot$$

Alkyl   Alkyl
Hydroperoxide   Radical

b. Homolytic Cleavage

Organic peroxides are very reactive and unstable. The oxygen - oxygen bond readily cleaves to yield two radical species which further react to form a wide variety of compounds such as ethers, esters, alcohols and organic acids.

$$-\overset{\backslash}{\underset{/}{C}}\text{-O-O-H} \;\rightarrow\; -\overset{\backslash}{\underset{/}{C}}\text{-O}\cdot \;+\; \cdot\text{O-H}$$

2. Olefins

Crude oil typically contains little to no olefinic compounds. Through refining and processing, however, olefins are produced and become a part of various crude oil fractions. Olefins can be found in thermally cracked and catalytically cracked gasoline fractions as well as in FCC cycle oils and coker gas oils. For this reason, it is not unusual for finished gasoline and distillate blends to contain a high olefin content stream.

A variety of fuel performance problems can be directly linked to the presence of olefinic compounds in a fuel. Problems such as darkening of fuel color, gum and sludge formation and combustion system deposits can be directly linked to the presence of olefins.

Double bonds between carbon atoms in fuel components are active sites for oxidative attack and polymerization. During long-term storage and during the combustion process, oxidation and polymerization of olefinic compounds can result in the following:

a. Varnish Formation on Metal Parts

The oxidation of olefins can result in the formation of organic hydroperoxides. These compounds readily decompose to form alcohols, carbonyl compounds and other oxidized species. These oxidized hydrocarbons can further react to form highly crosslinked, oxygen-rich materials. Some of these species can adhere to metal surfaces to form a hard "varnish-like" film or coating on metal parts. This "varnish" can act as a site for further deposition and eventual corrosion of metal. In severe cases, varnish can interfere with the hydrodynamic lubrication of moving metal parts and efficiency of component operation.

b. Drop-out of High Molecular Weight, Polymerized Fuel Components as Sludge

As fuel olefins polymerize, intermediate compounds form which often further react to form high molecular weight, fuel insoluble components. These high molecular weight materials can settle to the bottom of fuel storage tanks and transfer lines as sludge. Also, sludge deposits may act as initial sites for corrosion and microbial growth.

For example, the decomposition of a hydroperoxide to generate an alkoxy free radical can result in the reaction of the alkoxy radical with an olefin. A carbon radical then forms. Olefin chain propagation and polymerization can follow to yield high molecular weight deposits.

$$RO\cdot \;+\; \overset{\diagdown\;\diagup}{\underset{\diagup\;\diagdown}{C=C}} \;\rightarrow\; ROCH_2CH_2\cdot \;\rightarrow\; \text{Olefin Chain Propagation}$$

c. Color Bodies

Shifting and rearrangement of bonds within a molecule due to reaction with oxygen can result in the formation of compounds which impart color to fuel. These compounds often have some degree of aromatic functionality.

The color imparted to the fuel is typically amber to brown. Some distillate fuels turn deep red in appearance while others appear pale green.

As a general rule, a fuel component will impart color if it contains four or five conjugated *chromophoric* or *auxochromic* groups. Listed in **TABLE 4-8** are examples of common chromophoric and auxochromic groups.

TABLE 4-8. Common Chromophoric and Auxochromic Groups

CHROMOPHORE	*AUXOCHROME*
$\diagdown C = C \diagdown$	-OH
$\diagdown C = O$	RO-
-N=N-	$-NH_2$
$\diagdown S = O$	-SH
$\diagdown C = O$ | OH	$-NR_2$ -NHR

Fuel components possessing conjugated olefins, conjugated carbonyl compounds, or any combination of chromophores or auxochromes can be the source of color bodies in fuel. Examples are provided in **TABLE 4-9**.

TABLE 4-9. Examples of Color Bodies Identified in a High Sulfur Diesel Fuel

Compound	Description
C_1 to C_4 Indoles	White to yellow in color turning red on exposure to light
C_1 to C_4 Carbazoles	Exhibits fluorescence and long phosphorescence on exposure to UV light
C_1 to C_4 Phenols	Phenols turn pink or red if not perfectly pure or when exposed to light. This change is hastened in an alkaline environment
C_1 to C_2 Fluorenes	Fluorescent when impure

F. MICROORGANISMS, SEDIMENT AND WATER

Problems related to microorganisms, sediment and water in fuels and fuel systems can develop for a variety of reasons. Sources include:

1. Products of Microbial Metabolism

In fuel systems containing water, certain microorganisms can survive. Typical species are the bacterial anaerobe *Disulfovibrio* and the fungal species *Cladosporium resinae*. These microorganisms are called <u>H</u>ydrocarbon <u>U</u>tilizing <u>M</u>icrobes and often referred to as "HUM" or "HUM-bugs." They can initiate corrosion in fuel systems.

Fuel sulfur is reduced to H_2S by *Disulfovibrio*. Hydrogen sulfide can attack iron to form an iron sulfide scale and result in loss of iron from a metal surface. In the presence of water, iron sulfide scale can act as a site for severe pitting corrosion. This type of corrosion is often uncontrollable.

The term sulfate reducing bacteria (SRB) is frequently used to describe organisms which metabolize organic sulfates in fuel. Upon metabolism, the oxygen bound to the sulfate sulfur is consumed by the SRB and utilized in cellular respiration. The sulfur is reduced to H_2S gas.

Once liberated, H_2S can react with fuel olefins to form mercaptans, contribute to microbial induced corrosion or escape into the fuel.

Mercaptan Formation from Hydrogen Sulfide and an Olefin

$$\text{cyclopentylidene}=CH_2 + H_2S \xrightarrow{h\nu} \text{cyclopentyl}(H)-CH_2SH$$

Once microbes have multiplied and established themselves as a colony, they can adhere to metal parts forming "microbial plaques." Underneath these plaques, severe corrosion of metal is often found.

MICROBIOLOGICALLY INFLUENCED CORROSION (MIC)

Active microbiologically influenced corrosion (MIC) of metal due to bacteria is quite complex. It can involve several species of microorganisms and is affected by temperature, TOC, pH and other factors. Examples of bacterial species which are associated with MIC are described below:

Sulfate Reducing Bacteria (SRB)

Bacterial anaerobes including *Desulfovibrio*, *Desulfomonas* and *Desulfotomaculum* are known sulfate reducing species. They can survive in fresh, brackish and sea water and are present in most soils and sediments containing sulfate and sulfite compounds. Enzymes which promote the conversion of sulfates and sulfites to metal sulfides through chemical reduction are present in these bacteria. Iron sulfide, FeS, is a product of this process.

Acid Producing Bacteria

Organisms such as *Thiobacillus thiooxidans* and *Clostridium* species have been linked to accelerated corrosion of mild steel. Aerobic *Thiobacillus* oxidizes various sulfur containing compounds such as sulfides to sulfates. This process promotes a symbiotic relationship between *Thiobacillus* and sulfate reducing bacteria. Also, *Thiobacillus* produces sulfuric acid as a metabolic by-product of sulfide oxidation.

Clostridium species are anaerobic, spore-forming microbes. The formic, acetic, propionic and butyric acids produced as a result of their metabolic activity can enhance the corrosion of steel.

Metal Depositing Bacteria

Bacteria which oxidize ferrous iron (Fe^{++}) to ferric iron (Fe^{+++}) such as *Gallionella* and *Leptothrix* species are termed metal depositing bacteria. The result of this metabolic process is the formation of ferric hydroxide.

Slime Forming Bacteria

Most slime forming bacteria are aerobic. Species such as *Pseudomonas* produce an extracellular, gel-like, polysaccharide capsule which acts to protect and shield the organism. When in combination with other metabolic by-products, bacteria and water, *Pseudomonas* species form slime-like films and deposits on surfaces exposed to the air. Sulfate reducing species and acid formers can frequently be found in high concentrations beneath these slime layers. Also, slime layers can form masses large enough to plug filters and strainers.

Other Microbiological Factors

As bacteria die, ammonia is typically produced as a product of decomposition. Aerobic nitrifying bacteria such as *Nitrosomonas* and *Nitrobacter* can oxidize ammonia (NH_3) to nitrate (NO_3^-). As a result of this process, the pH of the system can be reduced.

Typical Microbiological Analysis in a Service Water System Pipe Experiencing Microbiologically Influenced Corrosion*

	Water	Deposits and corrosion products
Total aerobic bacteria	490,000	1,100,000
Enterobacter	30	<1000
Pigmented	<70	<1000
Mucoids	<10	<1000
Pseudomonas	100,000	210,000
Spores	12	8700
Total anaerobic bacteria	35	150,000
Sulfate reducers	30	120,000
Clostridia	5	30,000
Total fungi	2	<100
Yeasts	2	<100
Molds	None	<100
Iron-depositing organisms	None	None
Algae		
Filamentous	Very few	None
Nonfilamentous	Few	Few
Other organisms		
Protozoa	Few	None

*All counts expressed as colony-forming units per milliliter or gram.

2. Catalyst Fines

Residual fuel oils and heavy marine fuels are composed of high boiling petroleum fractions, gas oils and cracked components. Residual and clarified oil streams from the FCC process can contain degraded alumina/silica catalyst fines. These 20 - 70 micron diameter fines are known to contribute to a variety of problems in fuel injection and combustion systems. In marine engines, excessive injector pump wear, piston ring wear and cylinder wall wear can all be due to the abrasive action of catalyst fines on these fuel system parts.

BLUE SMOKE

One effect of valve sleeve, piston ring and cylinder wall wear is leakage of lubricating oil into the combustion chamber. When lubricating oil accumulates in the combustion chamber and burns with fuel, the exhaust smoke appears blue in color. Wear due to abrasion or corrosion has the same effect. At times, unburned fuel can also appear as blue smoke if fuel droplets are finely dispersed in the exhaust. Blue smoke is a symptom of a mechanical problem rather than a fuel performance problem.

3. Fuel Oxidation and Degradation Products

Certain fuel components, especially olefins, alkylated naphthenes and various heterocycles, can react with oxygen. Once oxidized, these compounds can further react to form fuel insoluble sludge and gums.

The term *varnish* is used to describe the hard, amber-colored coating of fuel oxidation products adhering to engine components. The term *sludge* is used to describe the heavy, dark-colored deposits which settle from solution out of the fuel. Sludge can accumulate in areas of low turbulence and act as a prime site for initiating corrosion.

If water is present in a fuel system, sludge and gums also lead to plugging of fuel filters. In contact with water, sludge and gums can aggregate into larger gel-like particles which are readily trapped by fuel filters. The aggregates can be as large as 100 microns in diameter and will not pass through standard 5 - 15-micron fuel filters. They will, however, flow through most higher porosity, 130-micron tank filters.

4. Engine Oil Contamination

If deposits found on filters, pump parts and engine systems contain high levels of **zinc, calcium, phosphorus and sulfur,** the possible source of the contamination is lubricating oil. These elements are part of the composition of the engine oil additive package. The elements and the compounds in which they can be found are listed in **TABLE 4-10.**

TABLE 4-10. Elements Found in Most Engine Oil Detergent Additive Packages

Element	Engine Oil Source
Zinc	Zinc Dithiophosphate - - Antiwear/antioxidant additive in most engine oils and industrial oils
Calcium	Overbased Calcium Sulfonate and Calcium Phenate - - Detergent/acid neutralizing additive
Phosphorus	Zinc Dithiophosphate or Phosphonate Compounds; Phosphorus is also found in gear oil additives
Sulfur	Zinc Dithiophosphate; Sulfonates; Sulfur is also found in gear oil additives

5. Wear Metals and Other Contaminants from Lubricant Systems

Oils used to lubricate machinery, motors, hydraulic systems and other mechanical devices can sometimes contaminate fuel systems. These oils often carry with them low levels of the metals which wear from the lubricated surfaces of the mechanical components. Some common wear metals and possible sources of origin are listed in **TABLE 4-11**.

TABLE 4-11. Possible Sources of Metals Found in Used Lubricants

Metal	Possible Source of Origin
Aluminum	Pistons, turbocharger/blower, pump vanes, thrust washers, bearings
Chromium	Compression rings, anti-friction bearings, shafts
Copper	Oil cooler tubes, bearings, bushings, thrust washers, valve guides, injector shields
Iron	Cylinders, liners, pistons, rings, valves, valve guides, gears, shafts, clutch plates, anti-friction bearings, rust
Lead	Bearings
Nickel	Valves, valve guides, anti-friction bearings
Silver	Silver solder, wrist pin bushings, anti-friction bearings
Tin	Bearings, plating

Other elements such as **boron and silicon** can be found in fuel and oil system deposits. They can originate from the following sources:

Silicon   Gaskets sealant compound, antifreeze, oil antifoam, sand and dirt
Boron   Antifreeze, lubricant additive

6. Sea Water Contamination

Water from the sea obviously contains high levels of sodium and chlorine. Sea water, though, differs from other forms of salt water and fresh water in the ratio of **magnesium:calcium.** In fresh water, the ratio of magnesium:calcium is from about 1:2 to 1:3. In sea water, the ratio can range from 3:1 to as high as 4:1. This difference helps to distinguish sea water from fresh water. **TABLE 4-12** provides information on the concentration of various components found in sea water.

TABLE 4-12. Concentration of Major Components Found in Sea Water

Component	Concentration, g/kg
Bicarbonate	0.142
Bromide	0.065
Calcium	0.400
Chloride	18.980
Magnesium	1.272
Potassium	0.380
Sodium	10.561
Sulfate	2.649
TOTAL	34.449

CHAPTER 5

Utilizing Physical and Chemical Property Measurements to Identify Sources of Fuel Problems

Petroleum product physical and chemical properties such as viscosity, aromatic content and distillation profile can provide a wealth of information about product quality and performance. The information provided in this section can be used to help identify how specific physical and chemical property measurements can be used to identify and solve fuel problems.

A. HIGH VISCOSITY

Problems associated with difficulty in filtering, mixing and pumping fuel can usually be linked to an increase in fuel viscosity. Some examples of fuel performance problems associated with high fuel viscosity are described below.

1. Inadequate Filling of Diesel Fuel Injection Fuel Pump

The barrel and plunger of a distributor type diesel fuel injection pump rapidly discharges and refills with fuel during normal operation. If fuel viscosity is high, fuel flow can be impaired enough to prevent adequate filling of the barrel. The result is poor fuel distribution to the cylinders and poor engine performance. Contamination of fuel with high viscosity products or operation at excessively low temperatures can increase the viscosity of diesel fuel and result in fuel pumping problems. Also, if diesel fuel viscosity at 0°F (-17.8°C) is greater than 45 cSt, fuel pumping problems are likely to occur.

Listed in **TABLE 5-1** are minimum and maximum ASTM viscosity limits for a variety of fuel oil grades.

TABLE 5-1. ASTM Viscosity Limits for Fuel Oils

Fuel Type	Min.Vis @40°C, cSt	Max.Vis @40°C, cSt	Other Viscosities
Kerosene	1.0	1.9	
Aviation Turbine Fuel			Max. 8.0 cSt @ -20°C
#1 Low Sulfur Diesel	1.3	2.4	
#1 Diesel Fuel Oil	1.3	2.4	
#1 Fuel Oil	1.3	2.1	
#2 Low Sulfur Diesel	1.9	4.1	
#2 Diesel Fuel Oil	1.9	4.1	
#2 Fuel Oil	1.9	3.4	
#4 Diesel Fuel Oil	5.5	24.0	
#4 Fuel Oil (light)	1.9	5.5	
#4 Fuel Oil	5.5	24.0	
#6 Fuel Oil			Min. 92 cSt @ 50°C Max. 638 cSt @ 50°C
#6 Fuel Oil			Min. 15 cSt @ 100°C Max. 50 cSt @ 100°C

2. Distortion of Diesel Fuel Injection Pump Parts

The tolerances and clearances between the moving parts of a high pressure fuel injection pump are quite small. Because of this, high fuel viscosity can result in a friction coefficient increase within a high pressure injection pump. Frictional energy can lead to excessive heat buildup and distortion of pump components. The Streibeck Curve in **FIGURE 5-1** demonstrates how an increase in fuel viscosity can have a dramatic effect on the friction coefficient of a moving system, especially under high load conditions.

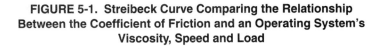

FIGURE 5-1. Streibeck Curve Comparing the Relationship Between the Coefficient of Friction and an Operating System's Viscosity, Speed and Load

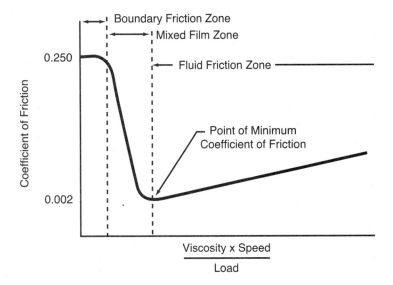

3. Poor Fuel Atomization

In order for fuel to combust and burn efficiently, it must be atomized into extremely small droplets. Fuel injectors aid tremendously in performing this role. However, if fuel viscosity is high, atomization into small droplets becomes difficult. Highly viscous fuel will not disperse freely after being sprayed from the fuel injector. The fuel instead sprays as a stream or large drops rather than as a fine mist. The result is a decrease in fuel efficiency and power due to incomplete burning of larger fuel droplets.

4. Low Flame Height in Wick-Fed Kerosene Burners

In order to maintain a constant feed rate of fuel to a burning flame, fuel viscosity must be low enough to travel effectively through the wick feeding a burning flame. The rate required to maintain an adequate supply of fuel to the flame can be affected by fuel viscosity. If viscosity is high, the rate at which fuel is supplied to the flame will be diminished. Consequently, the height of the flame is lessened due the lower amount of fuel feeding the flame. A secondary effect of using high viscosity fuel in wick-fed systems is a decrease in fuel consumption.

5. Pumping Problems with Residual Fuel and Marine Fuel Oils

The presence of high molecular weight compounds such as waxes, polycyclic aromatics, highly branched hydrocarbons and certain polymeric additives can increase the viscosity of a fuel. These compounds have a high resistance to flow. For example, a residual fuel oil with a viscosity greater than 1,100 cSt @ 40°C may be difficult to pump due to high percentages of asphaltic and paraffinic hydrocarbons.

High boiling point compounds such as asphaltenes, branched hydrocarbons with molecular weights greater than 300 and alkylated naphthenes can increase fuel viscosity. If present in fuel, they can be detected indirectly through either ASTM D-86 distillation end point temperatures or by high ASTM D-445 viscosity determinations.

Anytime residual fuel or marine fuel pumping problems occur, the first place to begin looking for the source of the problem would be the viscosity of the fuel. Pumps are designed and rated to move fluids within a fixed viscosity range. When fluids are either above or below the recommended viscosity range, pumping problems will eventually result.

B. LOW VISCOSITY

Low fuel viscosity can be due to the presence of low boiling, low molecular weight compounds in the fuel. Contamination with low boiling compounds such as solvents, gasoline and petroleum naphtha can dramatically reduce the viscosity of distillate fuel and residual fuel oil.

A reduction in both the initial boiling point and end point of a fuel can result in a lowering of viscosity. Accompanying a reduction in these distillation parameters is a consequential increase in the concentration of lower molecular weight and often, more volatile compounds.

The following performance problems can be due to low fuel viscosity:

1. Poor Penetration of Injected Diesel Fuel into Combustion Chamber

When diesel fuel is injected into the high pressure environment of a combustion chamber, it must fully penetrate into the pressurized air in order to completely combust. Fuel with a low viscosity does not have enough thrust to penetrate effectively through the pressurized air and disperse completely. The result is a soft, non-penetrating spray which does not spread throughout the chamber. Poor combustion efficiency and engine power result.

2. Wear of Fuel Injection System Components

Diesel fuel prevents wear of high pressure fuel pumping and injection equipment parts. If the viscosity of diesel fuel is low, its ability to form a hydrodynamic lubricating film between moving metal parts diminishes. The term "lubricity" is used to describe the wear inhibiting capability of distillate fuels.

If a diesel fuel is low in viscosity due to kerosene dilution or solvent dilution, its lubricity is probably poor. The possibility of wear of the high pressure fuel injection pump parts will increase. The effect of kerosene dilution on the lubricity of a typical low sulfur diesel fuel is shown in **FIGURE 5-2**.

FIGURE 5-2. Effect of Kerosene Dilution on Low Sulfur Diesel Fuel Lubricity

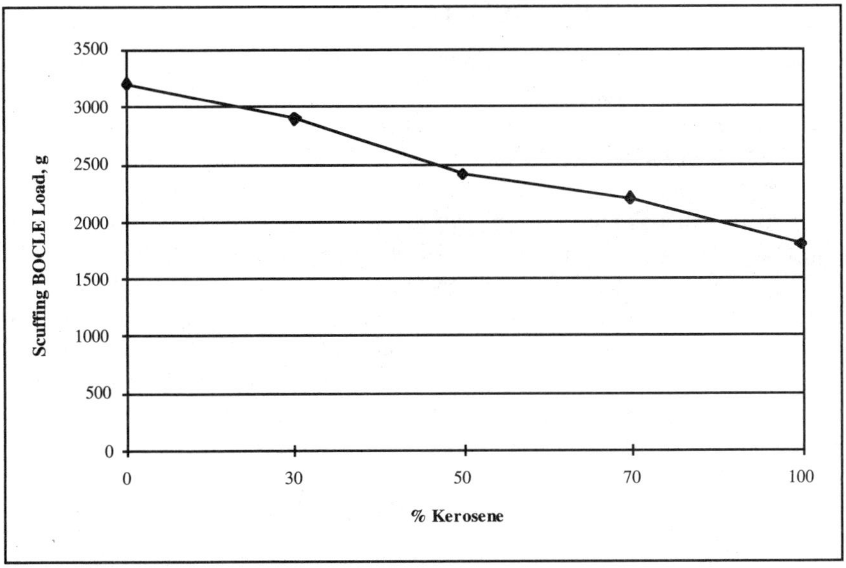

3. Hot Restart Problems in Diesel Engines

When fuel is moving through lines and engine parts which are hot, the fuel can "thin out" or behave as a less viscous material. This phenomena can lead to startability and engine operability problems in worn equipment.

For example, diesel powered vehicles operating under stop and go, high speed and load driving conditions can experience problems with restarting after the engine is turned off. This is due to the fact that fuel retained in the hot fuel injection pump leaks past the pump plunger and out of the injection barrel. Restarting the engine may be impossible until the engine cools down so fuel can again be held and maintained in the plunger and barrel assembly.

C. HIGH SULFUR CONTENT

Fuel sulfur content is being more closely scrutinized with each passing year. It is known that the burning of fuel sulfur can form sulfur dioxide and sulfur trioxide compounds. In combination with water, these sulfur oxides can form acidic compounds.

The following 11 problems have been associated with burning high sulfur content fuel:

1. Combustion System Corrosion

In fuel combustion systems, SO_2 and SO_3 can form upon the burning of fuel sulfur. When sulfur oxides combine with water vapor, acids form. This problem of acid formation and accumulation is a known phenomena and usually occurs under low speed and load operating conditions. The acids which condense on fuel system components can initiate corrosion of valves, piston rings and fuel injector nozzles.

In marine fuel applications, injector deposits and the corrosive wear of piston rings have been linked to fuel sulfur and sulfur bearing acid formation.

Engine lubricant formulators are aware of the problems associated with fuel sulfur and develop products to help combat its corrosive effect. Lubricants containing overbased calcium sulfonates and phenates are utilized to chemically neutralize the acids which form as a result of burning fuel sulfur. These oils can be effective at preventing the corrosive effects of fuel sulfur. However, the oils must be frequently changed to ensure that the acid neutralizing effect is maintained.

2. Copper, Bronze and Brass Corrosion

Sulfur readily attacks copper and its alloys, bronze and brass, to form copper sulfide, CuS. Elemental sulfur, mercaptan sulfur or hydrogen sulfide can all attack copper bearing parts. Fuel storage tanks and piping systems can contain copper heating coils, cooling coils as well as brass or bronze valves and fittings. These parts are all susceptible to sulfur initiated corrosion.

3. Exhaust Odor

The sulfur containing combustion gases sulfur dioxide and sulfur trioxide both have unpleasant odors. These gases form during the burning of fuel sulfur. Although less common, other sulfur containing compounds such as hydrogen sulfide, carbon disulfide or various mercaptans could form. All of these compounds have foul odors.

4. Possible Fuel Stability Problems

Substituted and condensed thiophenes and thiophenols are usually the most abundant sulfur containing compounds in refined fuels. These compounds are known to react with oxygen to form peroxides and eventually result in color bodies and gum-like fuel deposits. The reaction of thiophenol with a free radical compound and oxygen is shown below:

$$C_6H_5SH + R\cdot \xrightarrow{O_2} C_6H_5S\cdot + ROOH$$
Thiophenol / Hydroperoxide

$$C_6H_5S\cdot + C=C \rightarrow C_6H_5S\text{-}C\text{-}C\cdot \xrightarrow{O_2} C_6H_5S\text{-}C\text{-}C\text{-}O\text{-}O\cdot$$
Olefin

$$C_6H_5SH + C_6H_5S\text{-}C\text{-}COO\cdot \rightarrow C_6H_5S\cdot + C_6H_5S\text{-}C\text{-}COOH$$
Gum-Color Body Precursor

5. Elastomer Degradation

Mercaptan sulfur can degrade almost any elastomer. In aviation jet fuel, mercaptan sulfur is limited to 50 ppm.

6. Improved Diesel Fuel Lubricity Properties

Since the introduction of low sulfur diesel fuel, much study has been completed to determine the lubricity properties of this fuel. Comparison of low sulfur diesel with high sulfur diesel has clearly revealed that fuel sulfur has a dramatic impact on the ability of fuel to provide a higher level of lubricity performance. A comparison of the lubricity performance of a typical *high sulfur diesel, low sulfur diesel and a low aromatic low sulfur diesel* are shown in **FIGURE 5-3**.

FIGURE 5-3. Comparison of the Lubricity Performance of Typical High Sulfur Diesel, Low Sulfur Diesel and Low Aromatic Low Sulfur Diesel Fuel

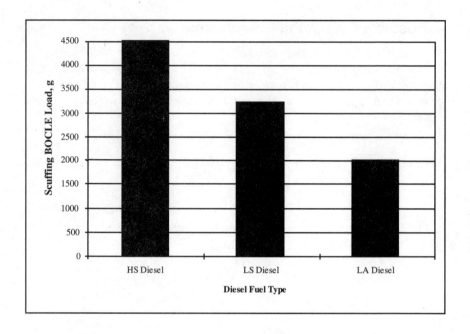

7. Combustion Chamber Deposits

Under low temperature, low speed and low load driving conditions, the fuel combustion system does not operate optimally. These conditions are commonly encountered during the winter months in stop-and-go city driving. Whenever fuel burns under these conditions, the water of combustion does not exhaust completely from the fuel system and condenses as fuel system moisture. Also, the SO_x formed upon burning and oxidation of fuel sulfur can be retained within the combustion system.

Together, water and SO_x can combine to form sulfur bearing acids. These acids can accumulate to initiate corrosive wear, oxidation of lubricating oil and the formation of piston lacquer deposits within the combustion chamber. Engine deposits can result in operability problems such as preignition knock, dieseling and wear.

8. Mercaptan Sulfur

Mercaptans or thiols are well known for their unpleasant odors. Although they are present in fuel at low levels, if at all, the role they play is significant. In addition to imparting unpleasant odor to fuel, mercaptans can attack copper and its alloys and can degrade fuel system elastomers.

When exposed to air, mercaptans are oxidized to yield disulfides and sulfonic acids.

$$\text{R-CH}_2\text{SH} \xrightarrow{[O]} \text{R-CH}_2\text{SSCH}_2\text{-R} \xrightarrow{[O]} \text{R-CH}_2\text{SO}_3\text{H}$$

Under extreme conditions of heat and in an acidic environment, aromatic sulfonic acids can revert to sulfuric acid and an aromatic compound.

$$\text{ArSO}_3\text{H} \xrightarrow[\Delta]{\text{H+}} \text{ArH} + \text{H}_2\text{SO}_4$$

Problems related to acid buildup include corrosion of metal system parts and degradation of the stability of fuel.

9. Elemental or "Active" Sulfur

Although rarely found in finished fuels, elemental sulfur could appear as a reaction product in fuels exposed to oxygen, rust or acids. The source of elemental sulfur in fuel is usually hydrogen sulfide gas. Reactions which can liberate sulfur can proceed as follows:

REACTION WITH OXYGEN

$$2\,\text{H}_2\text{S} + \text{O}_2 \rightarrow 2\,\text{S} + 2\,\text{H}_2\text{O}$$

AIR SPARGING OR REACTION OF DISSOLVED HYDROGEN SULFIDE WITH ATMOSPHERIC OXYGEN COULD YIELD ELEMENTAL SULFUR.

REACTION WITH RUST

$$3\,\text{H}_2\text{S} + \text{Fe}_2\text{O}_3 \cdot \text{H}_2\text{O} \rightarrow \text{S} + 4\,\text{H}_2\text{O} + 2\,\text{FeS}$$

RUST AND CORROSION PRODUCTS PRESENT IN STORAGE AND TRANSPORTATION EQUIPMENT COULD YIELD ELEMENTAL SULFUR.

REACTION WITH SULFURIC ACID

$$3\,\text{H}_2\text{S} + \text{H}_2\text{SO}_4 \rightarrow 4\,\text{S} + 4\,\text{H}_2\text{O}$$

ALTHOUGH INFREQUENT LOW CONCENTRATIONS OF SULFURIC ACID COULD CARRY OVER FROM ACID PROCESSING OR ALKYLATION.

Elemental sulfur is sometimes termed "active" sulfur because of its highly aggressive and corrosive nature toward metal, especially copper, bronze and brass. Copper sulfide readily forms in contact with elemental sulfur.

10. Degradation of Gasoline Engine Catalytic Converter Efficiency

Fuel sulfur and phosphorus have been identified in contributing to problems with catalytic converter efficiency. Fuel sulfur does poison converter catalysts, but not as severely as lead.

11. Diminishes the Effectiveness of Alkyl Lead Antiknock Compounds

Alkyl lead compounds are extremely effective gasoline antiknock agents. By decomposing to form lead oxide compounds during the gasoline combustion process, lead alkyls interrupt the rapid chain scission reactions which lead to combustion knock. Also, lead alkyls help to prevent exhaust valve seat wear and may minimize octane requirement increase. However, unless utilized in conjunction with lead scavengers such as 1,2-dichloromethane, lead deposits can accumulate within the gasoline combustion chamber.

Although being regulated out of use due to its toxicity and exhaust emission contribution, alkyl leads and other organometallics are still used in some parts of the world to improve automotive gasoline antiknock performance. Alkyl lead is also used in aviation gasoline.

Fuel sulfur can decompose alkyl lead compounds and can diminish the effectiveness of lead compounds at inhibiting combustion knock. The effect of fuel sulfur at degrading alkyl lead compounds can be rated as follows:

MOST EFFECTIVE	Polysulfides
	Thiols
↓	Alkyl disulfides
	Alkyl sulfides
	Elemental sulfur
↓	Aryl disulfides
	Aryl sulfides
LEAST EFFECTIVE	Thiophenes

D. HIGH AROMATIC CONTENT

Monoaromatic, diaromatic and polycyclic aromatic compounds can all be found in fuels. Although concentrations vary in different fuels, the presence of these aromatics can have both desirable and undesirable consequences. Examples include the following:

1. Low Temperature Flow Properties

As a general rule, low molecular weight aromatic compounds help to improve the low temperature flow properties of fuels and oils.

Since highly aromatic fuels have little wax, they possess better natural low temperature handling properties than paraffinic fuels. Also, the cloud point, pour point and low temperature filtration of aromatic diesel fuel will typically be much lower than a paraffinic diesel fuel.

2. Poor Cetane Engine Performance

Aromatic hydrocarbons do not have good performance in the cetane engine. Compared to paraffins of the same carbon number, aromatic compounds have a very low cetane number.

If a diesel fuel is reported to have a lower than expected cetane number, the first place to look for an explanation would be the total aromatic content of the fuel. A comparison of the approximate cetane number of various pure hydrocarbons is provided in **TABLE 5-2**.

TABLE 5-2. Approximate Cetane Number of Pure Hydrocarbons

Component Description	Number of Carbons	Cetane Number
n-Paraffin	10	82
n-Olefin	10	68
Alkyl Cyclohexane	10	50
Alkyl Benzene	10	<20
n-Paraffin	16	100
n-Olefin	16	95
Alkyl Cyclohexane	16	80
Alkyl Benzene	16	55
Alkyl Naphthalene	16	32

3. High BTU Value

The heating value of a fuel, often referred to as the calorific value or BTU value, can be directly linked to the aromatic content of the fuel. Since aromatic molecules are more tightly compacted than linear or branched paraffins, it is possible to "pack" more of them into the same volume or space.

To illustrate this, one quart of xylene, an eight carbon aromatic compound weighs approximately 823 g (1.81 lbs.) while one quart of n-octane, an eight carbon paraffin, weighs approximately 665 g (1.47 lbs.).

Fuel sold as heating oil typically has a higher aromatic content than fuel sold for use in an internal combustion engine. The higher aromatic content signifies that more carbon atoms are present per volume of fuel than fuel with a lower aromatic content. Therefore, more carbon is present to burn per unit volume. As a result, more potential heat is available from aromatic heating oils.

Volumetric Fuel Consumption

Studies documented by the Council of the Institution of Mechanical Engineers have compared fuel density with volumetric fuel consumption. Test results have shown that there is a general trend in both DI and IDI engines toward lower volumetric fuel consumption whenever higher density fuel is utilized. The results in **FIGURE 5-4** illustrate the relationship between fuel density and the volume of fuel consumed per horsepower-hour. The results shown in FIGURE 5-4 were obtained under full load engine operating conditions.

FIGURE 5-4. Relationship Between Fuel Density and Volumetric Fuel Consumption in DI and IDI Engines

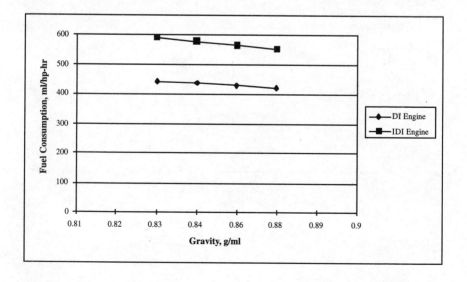

4. High Specific Gravity/Low °API

The specific gravity of a fuel or oil is a function of the weight per standard volume of the product. Since aromatic compounds have a greater weight per unit volume than do paraffinic hydrocarbons, the specific gravity of a highly aromatic product would be greater than a paraffinic fuel or oil.

The American Petroleum Institute (API) established an arbitrary rating system to measure the specific gravity of crude oils and petroleum products. This system assigns water the °API value of 10. Products heavier than water have °API gravity values <10. Most petroleum products are lighter in weight than water and have °API gravity values >10. For example, diesel fuel has a typical °API of 30 - 40 and gasoline has a typical °API of 50 - 60. Some asphaltic oils are heavier than water and have °API values <10.

Because fuel and oil weights vary per standard volume, they should always be handled and sold on a weight basis, not volume basis. If sold on a volume basis, the volume should always be corrected to 60°F (15.5°C) by utilizing °API volume reduction tables. These tables are used to standardize all volumes of petroleum products to a 60°F (15.5°C) volume. Standard °API volume correction tables for petroleum products are contained in **APPENDIX 3**.

5. Low Cetane Index

A widely accepted relationship exists between the cetane index of a distillate fuel and the total aromatic content. Typically, as fuel aromatic content increases, the fuel cetane index decreases. It is generally believed that distillate fuel containing >35% aromatics will have a cetane index <40. Likewise, fuel having <35% aromatics will have a cetane index >40.

6. Gum Formation

Heterocyclic aromatics in fuel such as pyridine, indole and condensed thiophene compounds are known to darken fuel color. They have also been shown to lead to an increase in the deposit forming tendencies of fuel. Aromatic peroxides can also react to form higher molecular weight, sludge-like material in fuel.

If highly oxygen rich, these deposits can adhere to metal parts in fuel systems. These adherent deposits can cause operating problems in engine components where tolerances are tight. Fuel pump plunger and valve guide tolerances are all quite small. Sticking of any of these parts due to the accumulation of adherent gum can result in poor operation and even damage of the moving parts.

7. Seal Swelling

Certain elastomeric materials used as gaskets, seals and hoses can be degraded by aromatic fuel components. Elastomers such as natural rubber, neoprene, Buna-N and ethylene-propylene will all swell in the presence of aromatic fuel components.

Sometimes, swelling is desirable if the purpose of the elastomer is to help provide a tight seal between parts which have been bolted together or mounted onto an engine. Swelling helps to prevent fuel leakage or air aspiration problems. As an example, swelling of seals used in the assembly of high pressure fuel pumps is desirable. Swelling helps prevent fuel leakage.

Degradation of elastomers used in transfer lines and hoses by aromatic fuel components is undesirable. For this reason, elastomers used in fuel storage and distribution systems are often made of hydrocarbon resistant elastomers such as Viton, PVC and Teflon.

8. High Solvency Power/High KB Value

The solvency of a petroleum product is a function of the aromatic content. The Kauri-Butanol Value (KB Value) is used to describe the "solvency power" of a petroleum product. There are both advantages and disadvantages associated with high fuel or oil KB Values.

For example, if a fuel with a high KB value were to be introduced into a system containing sludge or gum-like deposits, the fuel may act to dissolve and solubilize some of these deposits. As a result of this, the fuel may appear dark in color because it now contains dissolved degradation products. This fuel could then lead to tank filter plugging problems due to the trapping of dissolved sludge by the fuel tank filter.

The fact that naphthenic lubricants have a higher "solvency power" than paraffinic lubricants is used to an advantage in the formulation of certain oils. Originally, lubricants used in railroad engines were all naphthenic based. Naphthenic oil is effective at solubilizing deposits formed within the engine crankcase, thus preventing the settling and accumulation of deposits.

Another example would be the use of naphthenic oils in the formulation of compressor lubricants, textile processing oils and other industrial lubricants. In these formulations, solubilizing and dispersing compounds within an oil is quite important.

9. Asphaltene Incompatibility/Dropout in Marine and Residual Fuel Oils

Heavy residual components used in marine fuel applications typically contain significantly high levels of asphaltenes. When marine fuels of widely different viscosities are blended together, the asphaltenes may drop out of the final fuel blend. This phenomenon would be more pronounced if a low viscosity, paraffinic fuel were to be blended with a heavy residual fuel containing visbreaker bottoms. Consequences of asphaltene dropout could include excessive centrifuge sludge accumulation, injection system deposit formation and exhaust port fouling. Power loss could result from deposit formation and fouling.

10. Diaromatics and Diesel Fuel Lubricity

Recent studies have shown that alkylated diaromatic and polyaromatic compounds can have a positive effect toward improving the lubricity of fuel. These high boiling, high molecular weight species, such as alkylnaphthalene, can boost lubricity by several hundred grams when measured by the Scuffing BOCLE Test.

11. Black Smoke in Diesel Engine Exhaust

Black smoke is typically produced in diesel engines operating at or near full load. This occurs because the fuel volume injected exceeds the volume of air available in the combustion chamber needed for complete combustion of fuel carbon. Carbon particles and soot form to give the smoke a black or dark gray color.

Highly aromatic, low API diesel fuel can also enhance black smoke production. This is due to the fact that more fuel mass is injected per volume of fuel than would be from, higher API fuel. Also, there are a variety of mechanical factors which can influence black smoke production such as poorly functioning injectors and air intake system problems.

E. HIGH PARAFFIN CONTENT

Linear, branched and cyclic paraffins all exist in refined fuel. Fuel performance problems can often be directly related to the type and concentration of paraffin present. **TABLE 5-3** provides information on the typical carbon number range and boiling point temperatures of paraffins found in several representative fuels and other petroleum products.

TABLE 5-3. Typical Carbon Number Range and Boiling Point Temperatures of Paraffins Found in Fuels and Other Petroleum Products

Fuel	Paraffin Carbon Number Range	Paraffin Boiling Temperatures, °F (°C)
Gasoline	4 -12	31 - 421 (-0.5 - 216)
Mineral Spirits	9 - 12	303 - 421 (151 - 216)
Kerosene/Jet Fuel	9 - 16	303 - 548 (151 -287)
Diesel Fuel	9- 17	303 - 575 (151 - 302)
#6 Fuel Oil	12 - >20	421 - >650 (216 - >343)
Wax	18 and greater	601 and greater

Some of the important fuel characteristics associated with high fuel paraffin content are described as follows:

1. Poor Solvent/Low KB Value

Paraffins function poorly as a solvent for some organic compounds. This fact can have various consequences. For example, gums, deposits and fuel degradation products will not be dissolved or held in solution by high paraffin content fuels. As a result, gums and degradation products will fall from solution and settle onto fuel system parts such as storage tank bottoms and fuel system lines. The KB value for selected petroleum products is provided in **TABLE 5-4**.

TABLE 5-4. KB Value of Selected Petroleum Products

Product	KB Value	% Aromatics
Kerosene	30	<25
Hydrotreated kerosene	27	<1
Xylene	98	99+
Heavy Aromatic Naphtha	105	86+
Mineral Spirits (Stoddard)	39	15
140-Solvent	35	15
VM&P Naphtha	38	10
Heptane	38	<3

Also, additives which are used to enhance the performance of fuels and oils are usually highly polar organic compounds. Their solubility in paraffinic systems can be quite low unless formulated with paraffinic side chains or appropriate cosolvents. Additive dropout and deposition can occur unless paraffin solubility is maintained.

2. Diesel Fuel Cetane Number Increase

In general, paraffinic hydrocarbons have a higher engine cetane number than aromatic compounds with the same number of carbon atoms. Also, linear paraffins and alkenes have higher cetane numbers than cycloparaffins of the same carbon number.

For this reason, fuels refined from paraffinic crude oils will have naturally higher cetane numbers than fuels refined from naphthenic or more aromatic crude oil. Some typical cetane number values for selected n-paraffins, olefins, naphthenes and aromatics are shown in **TABLE 5-5**.

TABLE 5-5. Approximate Cetane Number of Various Hydrocarbons

Compound	Type	Number of Carbons	Cetane Number
Octane	n-paraffin	8	64
1-Octene	olefin	8	41
Decane	n-paraffin	10	77
1-Decene	olefin	10	60
Decahydronaphthalene	naphthene	10	42
n-Nonylbenzene	aromatic	15	50
Hexadecane (n-Cetane)	paraffin	16	100
1-Hexadecene	olefin	16	84

3. Diesel Fuel Wax Related Problems

At low temperatures, usually less than +20°F (-6.7°C), waxy paraffins may begin to crystallize in highly paraffinic diesel fuel. These paraffin wax crystals can cause a variety of fuel handling and performance problems. Examples include:

- Plugging of small lines and orifices
- Accumulation on fuel filter media to slow and eventually halt fuel flow
- Pumping difficulty due to an increase in the viscosity of the fuel
- Wax "dropout" or settling from fuel during long term storage. The result is accumulation of a wax layer at the bottom of a storage tank. This accumulated wax can cause line and filter plugging when pumped from the tank bottom and into vehicles.

In cold temperature environments and during winter months, fuel refiners and marketers will often blend kerosene or #1 diesel fuel into #2 diesel fuel to help combat the problems associated with paraffin wax in fuel. This technique is effective to a limited extent due to the reduction in fuel BTU and specific gravity values by these lighter fuels.

The paraffins in diesel fuel can vary in molecular weight, concentration and crystallization temperature. For this reason, fuels with similar physical properties may behave differently at low temperatures. This difference in behavior can be related to the low temperature properties of the fuel paraffins present.

4. Less Smoke During Burning and Combustion

Compared to highly aromatic fuels, paraffinic fuels combust and burn with less smoke and soot. This is primarily due to the fact that paraffins are hydrogen rich compared to aromatics. The hydrogen rich nature of paraffins enables the combustion process to progress more completely to CO_2 and H_2O. This is demonstrated by the following reaction equations:

Paraffin Carbon Combustion

$$-CH_2- + 1.5\ O_2 \xrightarrow{\Delta} H_2O + CO_2$$

Aromatic Carbon Combustion

$$2\ -CH- + 1.5\ O_2 \xrightarrow{\Delta} H_2O + CO_2 + C^o$$

As demonstrated, *two* aromatic carbon atoms must be burned to yield enough hydrogen to produce one water molecule. This process yields two carbon atoms which must be oxidized to either CO or CO_2. Without excess oxygen, free carbon-carbon reactions can result to produce amorphous soot and smoke. By increasing the amount of oxygen delivered into the fuel combustion process, the complete oxidation of all aromatic carbon to either CO or CO_2 is possible.

Paraffin combustion, however, can proceed more completely to H_2O and CO_2 because each carbon atom carries with it at least two hydrogen atoms. By burning only *one* paraffin carbon atom, enough hydrogen is released to react with oxygen to produce a molecule of water. The paraffin carbon can further react to form CO or CO_2.

5. Gasoline Engine Knock

Gasoline containing high concentrations of linear paraffinic compounds will *not* have the ability to prevent knocking as effectively as more aromatic gasoline. Aromatic compounds have a longer ignition delay period than most paraffins and, therefore, can inhibit engine knock more effectively.

Knock can occur over the entire range of engine speed. During acceleration, a large volume of fuel enters the intake manifold and mixes with air. Vaporization and evaporation of the more volatile, low octane compounds occurs readily. These compounds enter the combustion chamber and detonate quickly to initiate knock. The higher octane, less volatile compounds do not vaporize as rapidly from the fuel volume.

Deposit formation within the combustion chamber can also lead to knock. Knock is explained by the following mechanisms:

INCREASE IN CYLINDER PRESSURE

The deposit mass acts to reduce the physical volume of the combustion chamber through air displacement. As a result, during the compression-ignition cycle the cylinder air volume is compressed to a higher pressure. When ignition occurs, engine knock will be enhanced due to higher cylinder pressures.

INCREASE IN CYLINDER TEMPERATURE

Engine cylinder deposits retain heat and inhibit cooling of the combustion chamber. Under high temperature combustion chamber conditions, volatile fuel components will ignite and begin to combust before spark-ignition begins. This premature ignition enhanced by hot deposits can lead to knock.

Additionally, if the engine cooling system fails to effectively remove heat from the combustion chamber, knock may result. Also, at high speeds, knock can also occur, but may go unnoticed due to mechanical noise from the engine.

6. Shrinking of Rubber Seals/Fuel System Leaking

Aromatic compounds are used as plasticizers and components in the processing of certain rubber products. Rubber products, such as nitrile rubber, are used to manufacture fuel system seals. Conventional diesel fuels containing aromatic compounds will act to swell these seals and prevent fuel system leakage.

When highly paraffinic, low aromatic fuels are used in these same systems, aromatic components can leach from the seals and cause seals to shrink. As a result, fuel begins to leak from the fuel intake and distribution system.

F. LOW FLASH POINT

Flash point is considered to be an important specification for all finished fuels and oils. The flammability and combustibility characteristics of a material are directly related to the flash point. Also, fuel transportation codes require flammable compounds to be appropriately labeled for safety reasons.

According to OSHA, compounds with flash point values <100°F (37.8°C) are considered flammable. DOT and UN codes rate compounds flammable when the flash point is <141°F (60.5°C).

The following performance conditions could be attributed to a low flash point:

1. Low Diesel Autoignition Temperature

The presence of low molecular weight, low flash point compounds in diesel fuel could lead to a shortening of the fuel ignition delay period. In a diesel engine, this could cause rough running due to early combustion of the low flash point compounds.

2. High Reid Vapor Pressure of Gasoline

Vaporization of low boiling point, low flash point fuel components can increase the vapor pressure of a fuel. For this reason, fuel vapors are being closely monitored and regulated due to their tendency to escape into the atmosphere from fuel storage and distribution systems. Compounds with low flash points can contribute significantly to increasing fuel vapor pressure.

3. Vapor Locking in Gasoline Engines

Low boiling point compounds in fuel are necessary to ensure good starting and warmup of engines. During winter months most fuel refiners formulate fuel to provide good cold weather startability. Examples include increasing the RVP of gasoline in winter months and reducing the IBP of distillate fuel in cold operating environments.

However, under these circumstances, vapor locking problems can result if atmospheric or engine operating temperatures increase unexpectedly. Vapor lock occurs most frequently in engines with fuel pumps or fuel lines located near hot engine parts. When hot engines are turned off, fuel held in the fuel pump and in lines feeding the fuel pump can boil or vaporize. Also, vapors in fuel lines can push fuel away from the fuel pump. Upon restarting, fuel lines void of fuel will "lock" because the fuel pump cannot effectively pump the vaporized fuel.

Vapor locking problems can often be solved by either allowing the engine to cool or by pouring cold water over the hot fuel pump and lines. Both of these measures will allow fuel vapors to condense.

G. HIGH CARBON VALUES FOR MICRO METHOD, CONRADSON AND RAMSBOTTOM CARBON NUMBER DETERMINATIONS

The amount of carbon present in fuel components can be correlated with a tendency to form deposits in fuel systems. Although the use of various detergent and dispersant additives helps to minimize deposit formation, the carbon residue value is still quite useful.

The following information can be obtained from Micro Method, Conradson and Ramsbottom carbon values:

1. Contamination with Residual Fuel

A high carbon value for gasoline, jet fuel or #2 fuel oil is a good indication that the fuel has been contaminated with residual fuel oil. Heavy streams such as VGO, coker gas oil and #6 fuel oil can contaminate gasoline, jet fuel and diesel fuel. These streams tend to form carbon residue when pyrolized and can be identified as fuel contaminants through carbon residue testing.

2. Diesel Fuel Contains Alkyl Nitrate Cetane Improver

Diesel fuel containing cetane improver generates higher levels of residual carbon than untreated fuel. This is probably due to the fact that the cetane improver decomposes to catalyze fuel polymerization and gum formation during fuel pyrolysis.

The effect of increased levels of cetane improver on the Conradson Carbon value of a typical low sulfur diesel fuel is illustrated in **TABLE 5-6.**

TABLE 5-6. Effect of Octylnitrate Cetane Improver on the Conradson Carbon Number of a Typical Low Sulfur Diesel Fuel

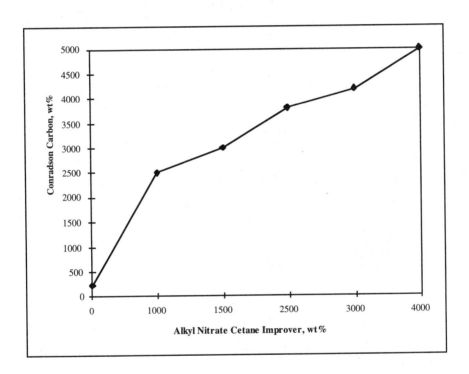

3. Fouling in Marine Fuel Exhaust Systems

High carbon residue values for marine diesel fuel, marine gas oil and heavy marine bunker fuel can contribute significantly to exhaust system deposit problems. Deposit formation on exhaust ports and exhaust turbines have been linked directly to high fuel carbon residue values.

H. DISTILLATION PROFILE

One of the easiest ways to determine the character and sometimes the quality of a fuel is to examine the distillation profile. Certain characteristics of fuel performance and potential fuel problems can be determined from the distillation profile of the fuel. Some important and known examples are provided as follows:

1. Smoke and Odor During Diesel Fuel Combustion

Two factors important to consider when identifying and solving fuel combustion problems are the amount of carbon present and the amount of oxygen available.

It is well known that the boiling point of hydrocarbon molecules will generally increase as the number of carbon atoms increases. It is also known that in order for fuel to burn cleanly and completely, enough oxygen must be present to react with all carbon atoms to form either CO_2 or CO. Therefore, if a fuel contains a greater number of carbon atoms, more oxygen will be required to completely combust this fuel to CO_2 or CO.

Diesel fuel with a high T-50, for example >575°F/302°C, will tend to burn with more smoke, soot and hydrocarbon odor than fuel with a lower T-50. This is basically due to the incomplete combustion and oxidation of a great number of high boiling, high carbon content fuel components in a limited oxygen content environment. Soot from combustion can result in the formation of carbon deposits in piston ring grooves.

2. Poor BTU Content of Heating Oil

The heating value or BTU rating of fuel oil used to generate heat is related to the amount of carbon available in the fuel for burning. If the T-50 of heating oil is <450°F, the amount of carbon available to serve as an efficient energy source is less than a higher T-50 fuel. Problems related to poor energy efficiency of fuel oil could be traced back to the low carbon content evident by the T-50 temperature.

3. Diesel Engine Warmup Difficulty

In order for a cold diesel engine to warm up effectively, fuel components must be present which have the proper autoignition characteristics to maintain smooth combustion. If a fuel contains a large proportion of components with widely different ignitability characteristics, smooth warmup will be difficult to achieve.

The conditions which exist in a cold, low RPM engine environment require fuel which has a significant amount of material which will autoignite within a limited, lower temperature range. This can be determined by looking at the boiling characteristics of the fuel. If a wide temperature range exists between the T-10 and T-50 values, it could be determined that the fuel contains a large variety of different components which all have different autoignition characteristics. A T-50 in the range of 450°F to 535°F (232°C to 280°C) is desirable for most high speed diesel fuel applications.

4. Poor Diesel Engine Startability

Initial starting of a cold diesel engine can be related to the volatility of the fuel. The IBP of diesel fuel and the T-10 provide some information on the startability of a diesel engine. If the fuel T-10 is high, for example >350°F (176.7°C), startability problems may be experienced.

5. Crankcase Dilution of Motor Oil/Carbon Deposits in Combustion Chamber of Gasoline and Diesel Engines

If fuel contains a large amount of high boiling, high carbon content components, it is possible that some of these component will not combust. In both gasoline and diesel engines, fuels with high T-90 temperatures will burn incompletely.

As a result of this, unburned fuel can remain in the combustion chamber to be drawn into the crankcase with the oil lubricating the piston cylinder, or it can remain within the upper part of the cylinder to "carbonize" and form combustion chamber deposits. Piston ring and land groove deposits can also form.

Fuel Enters Combustion Chamber
↓
Incomplete Combustion/Oxidation of High Boiling Components Results
↓
Partially Combusted/Oxidized Fuel Remains as Liquid on Piston and Cylinder Walls
↓
Liquid Collects in Piston Grooves, Lands and Lubricating Oil
↓
Continued Oxidation Results in Deposit Formation

The long-term effects of utilizing high T-90 fuel will lead to noticeable engine operability problems and possible engine damage.

Refiners understand that gasoline will tend to cause dilution of engine oil if the 90% point is high. The 90% point distillation limits shown in **TABLE 5-7** have been established by ASTM to help control crankcase dilution in various outside operating temperature environments.

TABLE 5-7. ASTM T-90 Limits for Gasoline in Various Operating Temperature Environments

ASTM D-86 Distillation 90% Point, °F	Minimum Outside Temperature, °F
370	80
350	60
340	40
325	20
310	0
300	-20

For example, if the 90% distillation temperature rises above 340°F when the outside temperature is below 40°F, crankcase oil dilution with gasoline may increase.

6. Inaccurate Cetane Index Ratings

It is not possible to accurately determine the cetane index for many #1 diesel and kerosene type fuel. This is because the cetane index scale cannot be utilized accurately for fuels with end points <500°F (260.0°C).

7. Knock in Gasoline Engine

Problems related to engine knock can be readily diagnosed by measuring the octane number of gasoline. However, it is also possible to look at the distillation profile of gasoline and anticipate problems linked to engine knock.

a. High T-90

High molecular weight fuel components have a longer ignition delay than most lower molecular weight components. Therefore, components of the gasoline T-90 fraction will naturally have more of a resistance to knock than other fractions. By reducing the T-90 of gasoline, knock resistance is also reduced.

b. "Front-End" Volatility

Both MON and RON values of finished fuel blends are measured using the whole fuel, not individual fuel fractions. However, the octane number contribution of the various fuel fractions can influence the overall knock resistance. Finished fuels with identical RON values may contain fractions with either very similar or widely different octane number values.

For example, the T-10 through T-40 fractions of available gasolines may possess significantly different octane numbers. This difference may be manifested in various levels of knock resistance under low temperature engine starting and warmup operation.

During cold weather engine warmup and acceleration, volatile fuel components would tend to vaporize and combust more readily than the higher boiling, less volatile compounds. Therefore, in cold weather, fuels with low octane number T-10 through T-40 fractions may tend to knock more readily during initial warmup and acceleration than fuels possessing higher octane number T-10 through T-40 fractions. Although modern engine design has helped to minimize this problem, cold weather engine knocking problems still exist.

CHAPTER 6

Solving Fuel Problems by Using Chemical Additives

A wide range of fuel performance problems can often be solved by the use of chemical additives. Although in some circumstances the additives used may be no substitute for refining processes such as hydrotreating, caustic washing or distillation, they can provide an alternative to further processing.

Some chemical additives such as corrosion inhibitors, wax crystal modifiers, detergents and demulsifiers provide performance which is difficult to duplicate through refining without adversely affecting some other fuel property. Other additives such as metal chelators, fuel sweeteners, biocides, lubricity improvers, foam control agents and combustion enhancers can too be used to solve fuel performance problems.

Recent developments and concern over the control of fuel exhaust emissions have led to the increased use of combustion system detergents, oxygenates and cetane improvers in fuel.

Oxygenated blend components such as ethanol, methyl *t*-butyl ether (MTBE), ethyl *t*-butyl ether (ETBE) and *t*-amylmethyl ether (TAME) are also used to help limit the exhaust emissions from fuel.

Some of the most frequently utilized additives effective at solving fuel related problems are described within this chapter.

A. ANTIOXIDANTS

Over time, oxygen will react with fuel components to degrade the fuel into a viscous, sludge-like mass. Fuel components most susceptible to oxidation are olefins and alicyclic naphthenes. Paraffins and aromatics are less susceptible to attack but can eventually be consumed by the free radical chain reaction process.

Hindered phenol and phenylenediamine (PDA) compounds are commonly used and quite effective at preventing free radical oxidative degradation of fuel. They can be used in gasoline, kerosene, jet fuel and certain distillates and lubricants. Often, a synergistic effect can be obtained by using a combination of a hindered phenol and a phenylenediamine antioxidant in the same application.

1. Hindered Phenol Antioxidants

Hindered phenol compounds usually possess alkyl groups on ortho and para sites. The alkyl groups are typically t-butyl or methyl in functionality. The lower cost of hindered phenol antioxidants makes them attractive for use in fuel applications. In gasoline, hindered phenols are typically used at treat rates of 5-50 ppm. The limitations placed on jet fuel additives often control the rate at which phenolic antioxidants can be used.

When contaminated or held in storage at temperatures below 60°F (15.5°C), hindered phenols may tend to crystallize and fall from solution as relatively pure compounds. Often solvent dilution of hindered phenols can help to reduce the tendency of crystal dropout, but will not completely eliminate this problem.

2. Phenylenediamine Antioxidants

Phenylenediamine (PDA) antioxidants can be used when hindered phenols do not perform or whenever their cost performance overcomes that of hindered phenol compounds. Although phenylenediamine compounds can often be as much as four times the cost of a hindered phenol antioxidant, the performance provided by PDAs can frequently exceed that of a hindered phenol compound.

Phenylenediamines are relatively strong organic bases and can cause burns if left on the skin for more than a few minutes. Their dark color makes them undesirable for use in fuels or petroleum fractions which have Saybolt color specifications >+15. Because of their alkaline nature, they may also interact with any acidic compounds which may be present in the fuel or in tank bottom water. Under extreme conditions, this interaction could result in a reduction or loss in performance of the PDA antioxidant.

3. Organosulfur Compounds

Some organosulfur compounds can function as fuel antioxidants by acting to decompose hydroperoxides. Organosulfides are believed to react with hydroperoxides to form sulfoxides. The sulfoxides then further react with hydroperoxides to form other more acidic compounds. These newly formed acids continue the process of decomposing and reaction with hydroperoxides. Thus, organosulfur compounds function in the process oxidation inhibition through hydroperoxide decomposition. However, in most fuel applications, sulfur containing antioxidants are not utilized.

Fuel problems which can typically be solved by a fuel antioxidant
- Control of color change and darkening
- Retarding the formation of deposits and varnish
- Improvement in the resistance of fuel to oxidation

Fuel problems which CANNOT typically be solved by a fuel antioxidant
- Reversing of the effects of fuel degradation such as color change or deposit formation

Solving Fuel Problems by Using Chemical Additives   135

- Effectively controlling the fuel degradation process after it has progressed significantly
- Significantly improving jet fuel JFTOT tube and screen ratings
- Color and deposit formation in distillate fuel are sometimes difficult to control by an antioxidant. Often, other components are needed.
- Reduction in the existent gum level of gasoline

General characteristics of phenolic and PDA antioxidants are summarized in TABLE 6-1.

TABLE 6-1. Characteristics of Hindered Phenol and Phenylenediamine Antioxidants

Hindered Phenol	Phenylenediamine
Effective in most fuel at 5-50 ppm	Effective in most fuel at 1-20 ppm
Will not significantly degrade fuel color	May degrade Saybolt color of fuel
Lower cost/pound	High cost/pound
Can crystallize during storage, especially at temperatures below 60°F; dilutions are available to help lower crystallization temperatures	Can crystallize during storage; crystallization temperature can be controlled by alkyl group functionality; products are available with crystallization range between 60°F (15.5°C) and -40°F (-40.0°C)

FIGURE 6-1. 2,6 - Ditertiarybutyl phenol
A typical hindered phenol type fuel antioxidant

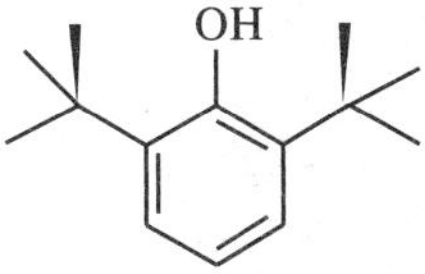

FIGURE 6-2. *N,N'*-Di-isopropyl phenylenediamine
A typical phenylenediamine antioxidant

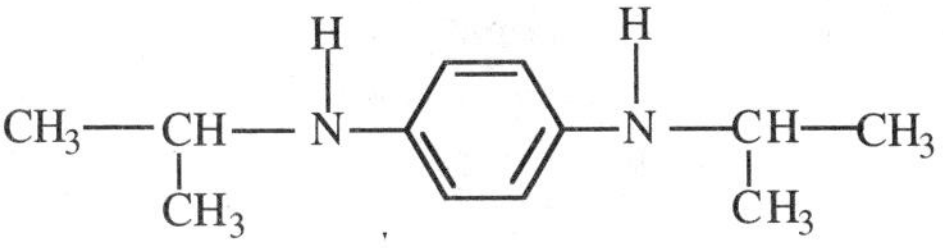

B. DISTILLATE FUEL STABILIZERS

Compounds classified as distillate fuel stabilizers are often mixtures of various compounds having different performance enhancing capabilities. These mixtures are frequently added as a single "package" to a distillate fuel. Stabilizer packages can possess any combination or all of the following components:
- Oxidation Inhibitor
- Dispersant
- Corrosion Inhibitor
- Metal Chelating Agent
- Acid Neutralizing Compound
- Microbiocide

Since distillate fuel contains a greater number of higher molecular weight compounds than those found in gasoline, the deposits which form can also be much greater in molecular weight. Heavy, sludge-like deposits and dark adherent gums are usually the products of distillate fuel degradation.

Naphthenic acids and sulfur and nitrogen containing heterocycles and naphthenoaromatic compounds can all be found in distillate fuel fractions. Degradation through condensation type reactions rather than through radical initiated polymerization can be more common in distillate fuel. Because of this, degradation products are often quite complex in nature and structurally diverse.

Refiners can utilize light cycle oil fractions and vacuum gas oils as distillate blending stocks. The percentage of cycle oil in a distillate fuel blend can range from 1% to 20%. Higher percentages can be used, but use is dependent upon the stability and properties of the cycle oil. Light cycle oils can contain unstable olefinic and naphthenic hydrocarbons. These compounds can react with oxygen to produce hydroperoxides and can also participate in condensation reactions. In either case, high molecular weight, deposit forming compounds result. Vacuum gas oils vary in composition and quality. Their effect on fuel stability can be a function of the nature of heterocyclic compounds present in the gas oil.

Traditional hindered phenol and phenylenediamine free-radical scavenger type antioxidants do not usually provide the stability performance desired in distillate fuel. Stabilizer formulations containing components which provide a wider range of performance are often required. Some of the components commonly used in distillate fuel stabilizers are described as follows:

1. Dispersant

Some distillate fuel stabilizers possess dispersant-like properties. By acting as a dispersant, any sludge or deposit-like component which may form can be suspended in the fuel and maintained as a soluble compound. As a result, deposits do not accumulate onto fuel system components, but remain dispersed in the fuel. However, due to this dispersing action, the fuel may appear dark in color.

Corrosion inhibiting properties of fuel stabilizers can be a secondary effect of the dispersing action of a fuel stabilizer. By functioning as a dispersant, sludge and water are held in suspension and prevented from initiating metal surface corrosion. Also, some stabilizer dispersants can form a thin film on the metal surfaces of fuel system components. This film-forming property enables the stabilizer-dispersant to function in corrosion control.

FIGURE 6-3. Example of an Oil Soluble Dispersant Compound

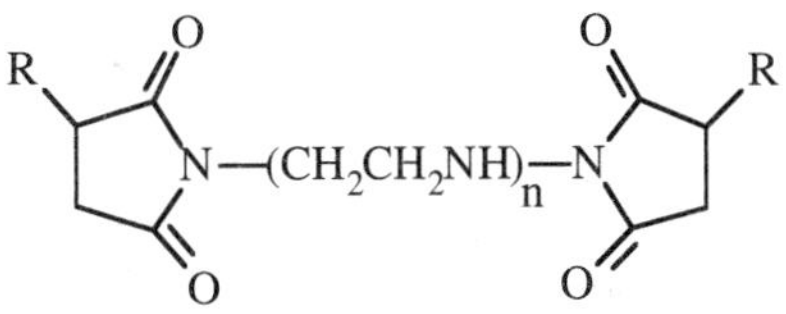

$$R \underset{O}{\overset{O}{\diagdown}} N-(CH_2CH_2NH)_n-N \underset{O}{\overset{O}{\diagup}} R$$

2. Metal Chelation

Mechanical components used in fuel systems such as pumps, valves and bearings may contain copper or copper containing alloys. As a fuel system component, copper is especially undesirable because it acts as a catalyst in promoting the oxidation of fuel paraffins to oxygen-rich, gum-like deposits. The following reaction sequence represents how copper ions can catalyze the oxidation and degradation of hydrocarbons:

Hydroperoxide Formation

$$Cu^{+2} + RH \rightarrow Cu^+ + H^+ + R\cdot \overset{O_2}{\rightarrow} Cu^{+2} + ROOH$$

Hydroperoxide Degradation

$$Cu^{+2} + ROOH \rightarrow Cu^+ + H^+ + ROO\cdot$$

$$Cu^+ + ROOH \rightarrow Cu^{+2} + OH^- + RO\cdot$$

The structural design of a fuel stabilizer molecule can enable some compounds to function in metal ion chelation. Most stabilizers have nitrogen atoms within the structure of the compound. The electron pair on nitrogen can function in metal chelation if the nitrogen atoms are oriented in the appropriate conformation.

Also, separate compounds known to perform as effective metal chelators can be blended into stabilizer formulations. Compounds such as N,N'-disalicylidene-1,2-propanediamine and N,N'-disalicylidene-1,2-cyclohexanediamine are commercially available and quite effective. Care must be taken when using these compounds in cold environments. These metal chelators can crystallize and precipitate as relatively pure compounds. At temperatures as high as 70°F (21.1°C), some metal deactivator formulations can form crystals. These crystals, however, readily go back into solution upon warming to about 90°F (32.2°C).

FIGURE 6-4. N,N'-Disalicylidene-1,2-propanediamine

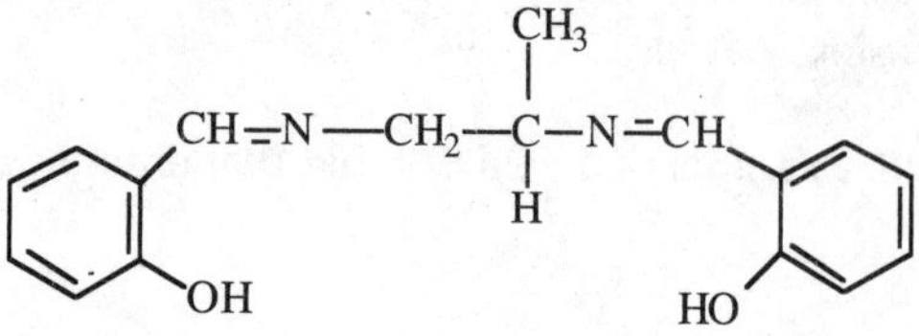

$$\text{HO-C}_6\text{H}_4\text{-CH=N-CH}_2\text{-CH(CH}_3\text{)-N=CH-C}_6\text{H}_4\text{-OH}$$

3. Antioxidant

The antioxidant property of most fuel stabilizer formulations helps to prevent the initial formation of most acids. Compounds such as tertiary-alkyl primary amines are quite effective distillate fuel antioxidants. These oil soluble amines can be formulated into stabilizer formulations at relatively low concentrations and can provide a significant boost in overall performance of a fuel stabilizer.

**FIGURE 6-5. 1,1,7,7-Tetramethyloctylamine
A Tertiary-Alkyl Primary Amine**

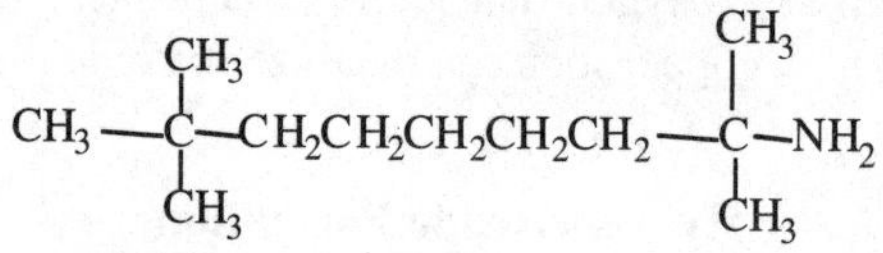

$$\text{CH}_3\text{-C(CH}_3\text{)}_2\text{-CH}_2\text{CH}_2\text{CH}_2\text{CH}_2\text{CH}_2\text{-C(CH}_3\text{)}_2\text{-NH}_2$$

FEDERAL DIESEL FUEL OIL STABILIZER

A diesel fuel stabilizer containing a blend of amines, polyamines and alkyl ammonium alkyl phosphate as specified under MIL-S-53021 is recommended for Federal diesel fuel oil. This stabilizer is to be used at a treat rate of 25 lb/1000 barrels. It is not intended for use in routine applications, but for situations where increased stability protection is required.

Typical applications include fueled equipment undergoing long-term storage in a warehouse or depot, pre-positioned equipment or equipment maintained in a high temperature environment.

NAVAL DISTILLATE STABILIZER

A fuel oil stabilizer additive listed in QPL-24682 may be blended into naval distillate fuel at rates up to 35 lb/1000 barrels to protect against degradation and improve storage stability as measured by ASTM D-5304. Method ASTM D-2274 may also be used if the test duration is extended from 16 hours to 40 hours.

Fuel problems which can typically be solved by a distillate fuel stabilizer

- Control of color darkening and sludge formation when cycle oil, cracked gas oil or some other unstable fraction is blended into fuel. These fractions contain unstable olefins and aromatics in high percentages.
- Control of color change and deposit buildup during the long-term storage of distillate fuel.
- Neutralization of naphthenic acids and other acidic compounds which may be present in fuel. Acidic organic compounds can react with other fuel components to initiate the formation of higher molecular weight, fuel insoluble deposits.
- Prevent copper catalyzed oxidation of fuel components. Copper ions can act as catalysts to initiate the rapid oxidation and degradation of fuel components. The result of this rapid oxidation will be a darkened, viscous fuel residue.

Fuel problems which CANNOT typically be solved by a distillate fuel stabilizer

- Change of fuel color from a dark ASTM color to a lighter ASTM color.
- Perform as an effective substitute for a hindered phenol or phenylenediamine antioxidant in inhibiting free radical reactions in gasoline or jet fuel.

C. DEMULSIFIERS AND DEHAZERS
1. Water and Contaminants

Water can contaminate fuel through several different routes. Examples include:
- Refinery processing and wash water
- Condensation of water in storage tanks vented to the atmosphere
- Barge ballast water carryover
- Storage tank water bottom carryover

Under normal circumstances, refined fuels do not form emulsions with water. The fuel and water readily separate into two distinct phases, a lower water phase and an upper fuel or oil phase. However, when emulsifying agents mix with fuel, emulsification can result. Examples of common fuel emulsifying agents include any of the following:

a. Microorganisms, Rust and Dirt

Tanks used to store fuel are usually emptied and filled on a continuous basis. Although most fuel refiners and marketers maintain some type of routine tank maintenance schedule, water and debris still accumulate.

Products of microbial metabolism, rust, salts of calcium, sodium, magnesium and iron as well as silicon in the form of dirt can all be found in tank bottom water. These materials help to stabilize a fuel-water emulsion.

b. Naphthenic Acids and Other Polar Compounds Naturally Occurring in Fuel

Distillate fuel fractions can contain naphthenic acids, sulfonic acids and other hydrophilic compounds. If these hydrophilic compounds are present as sodium salts due to caustic washing of fuel, they become powerful emulsifying agents. Also, heavy resinous compounds in fuel can act to stabilize existing emulsions.

c. Refining and Processing Additives

Certain corrosion inhibitors, antifoulants and other polar organic compounds used in the refining and processing of fuels may "carry over" into a finished product. If present at significant levels, these compounds can interact with water to form an emulsion. These emulsions are often quite difficult to break. The result is a cloudy, hazy fuel.

d. Detergents and Deposit Control Additives

Detergents added to gasoline and diesel fuel which help control the formation of deposits within engines are highly polar organic compounds such as succinimides and polyethers amines. When fuel containing a detergent mixes with water, emulsions can form which plug fuel filters and cause sticking of fuel injection pump parts.

The energy needed to initiate the formation of a fuel-water emulsion can come from sources such as a high speed centrifugal pump used to transfer products through pipelines or from the turbulence created during the loading of fuel into a storage tank or a barge hold contaminated with water.

2. Understanding the Terms Demulsifier and Dehazer

The terms *demulsifier* and *dehazer* are often used interchangeably to describe compounds which break fuel emulsions or remove water initiated haze from fuel. The terms can be distinguished as follows:

The process of *demulsification* usually involves removal or "dropout" of relatively large volumes of water which have contaminated fuel. The amount of water in this circumstance is usually >1% of the total fuel volume. Also, when an emulsion has been created in a fuel, breaking this emulsion is termed demulsification. Sometimes, the term *emulsion breaking* is also used to describe the process of demulsification.

Dehazing refers to removal of water which has been dispersed throughout a fuel matrix. Haze created by waterdroplets can be removed by coalescence or by further dispersion through the action of a chemical dehazer.

3. Demulsification and Dehazing Mechanisms

Two basic mechanisms have been proposed to explain the performance of demulsifiers and dehazers in finished fuels. These two mechanisms are defined as *coalescence and adsorption*. Both processes rely on the fact that the demulsifier contains both hydrophobic and hydrophilic sites. The mechanisms can be defined as follows:

a. Coalescence

Given time, water which exists as discrete droplets in finished fuel may coalesce into larger drops and settle by gravity from the fuel. Demulsifiers or dehazers can accelerate this process by functioning as a site for attraction of dispersed water.

Water is first attracted to the hydrophilic sites of the demulsifier through the process of hydrogen bonding. Eventually, enough water accumulates and coalesces into larger drops. The larger drops then separate from the hydrophilic sites and fall out of the fuel to the bottom of a storage tank or blending vessel.

b. Adsorption

If water is emulsified into fuel as a water-in-oil emulsion, coalescence cannot affect the removal of water from the fuel. The outer oil or surfactant layer surrounding water will not permit water to hydrogen bond to the hydrophilic demulsifier sites.

To break through this oil or surfactant layer and free the retained water, it is necessary to chemically disrupt the stability of this layer. Demulsifiers and dehazers can adsorb onto the protective film and subsequently interfere with the electrochemical forces which hold this outer layer together.

Upon adsorption, demulsifier/dehazer compounds function to break the oil or surfactant layer thus releasing the contained water. Once free, the water can then coalesce into larger drops and be removed from the fuel.

4. Preventing Emulsification

Many gasoline and diesel fuel formulations contain deposit control additives or detergents designed to keep fuel injection and intake systems clean. If water contacts fuel containing these detergents, emulsification of water with the fuel detergent may occur. The addition of demulsifying/dehazing compounds to fuel detergent formulations can prevent these emulsions from forming.

The mechanism of preventing detergent-water emulsions is controlled by the ability of the demulsifier/dehazer to interfere with the formation of a stable oil film around waterdroplets. The demulsifier/dehazer is believed to intermix with the fuel detergent to prevent the formation of this stable oil film. By preventing an oil film from forming, surrounding and encapsulating water, detergent enhanced emulsions can be inhibited.

5. Chemistry of Demulsifiers and Dehazers

Quaternary ammonium salts and salts of alkyl naphthalene sulfonic acid were some of the first compounds to be used effectively as fuel demulsifiers and dehazers. Today, a wide range of monomeric and polymeric demulsifiers and dehazers exist.

These compounds are typically produced by reacting an acceptor molecule with an epoxide. The acceptor can be monomeric or polymeric in nature and possess free hydroxyl groups onto which the epoxide reacts. The epoxide is typically ethylene oxide, propylene oxide, butylene oxide or a combination of oxides.

Sorbitol and glycerine are commonly used as monomers for oxide addition. Various alkyl phenol-formaldehyde compounds are examples of polymeric acceptor compounds having a large number of unreacted hydroxyl groups. The extent of oxide polymerization can have a significant impact on performance and solubility of the dehazer or demulsifier in fuel and oil systems.

As a general rule, the addition of ethylene oxide to a resin backbone will tend to increase the water solubility of the compound. The addition of propylene oxide or butylene oxide to the resin will tend to increase the hydrocarbon solubility of the compound. Often, the dehazer or demulsifier can be made to perform selectively in oil-water systems by adding both ethylene oxide and propylene oxide to the same molecule. Performance and solubility of the alkoxylated compound can then be finely tuned by closely controlling the amount and order of epoxide addition.

FIGURE 6-6. A Random EO-PO Based Fuel Demulsifier

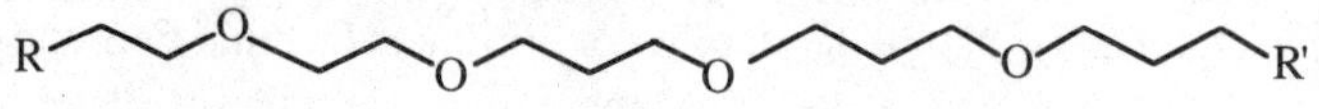

D. MICROBIOCIDES

The accumulation of water in the bottom of fuel storage tanks and fuel handling systems is a common problem throughout the fuel industry. Condensate water and process water in combination with the hydrocarbon fuel can serve as an acceptable medium to support the growth of various microorganisms. Anaerobic species such as *Clostridia*, *Desulfovibrio* and *Desulfomonas* can survive in the oxygen-free environment existing at the fuel-water interface of the storage tank. These microbes utilize the oxygen bound up in compounds for their metabolic oxygen needs.

Some bacteria specifically utilize oxygen bound in the sulfate complex of a compound. As a result of this metabolic activity, sulfur is reduced to H_2S. For this reason, these microbes are called "sulfate reducing bacteria" or SRBs. They can tolerate temperatures as high as 80°C (176°F) and environments from about pH 5 to pH 9. Species such as *Desulfivibrio* and *Desulfomonas* are examples of SRBs.

Many of these anaerobic, SRB species possess the ability to utilize hydrocarbon fuel compounds as a nutrition source. These species are termed "hydrocarbon utilizing microbes" or HUM. Fuel containing linear paraffins, especially distillate fuel, can be metabolized by these microbes through the β-oxidation process.

Solving Fuel Problems by Using Chemical Additives 143

By-products of anaerobic microbial metabolism include H_2S, H_2, CH_4 and low molecular weight organic acids. Also the formation of FeS and the accumulation of microbial cell decomposition products can complicate the contamination problem.

Colonies established by microbes on the sidewalls and bottoms of fuel storage tank systems are called "plaques." These plaques can often be the site of microbiologically influenced corrosion (MIC) of the underlying metal.

A variety of biocides are available for treating water contaminated with microorganisms. However, only those products which are EPA registered and approved for the specific fuel application can be utilized. For example, a biocide which is registered for use in treating only fuel oil storage tank water bottoms cannot be used to treat gasoline storage tank water bottoms. Before treating a fuel system with any biocide, the EPA registration must be confirmed.

Both water soluble and fuel soluble biocides are available for fuel treatment. The advantages and disadvantages of each method of treatment are briefly outlined in **TABLE 6-2.**

TABLE 6-2. Advantages and Disadvantages of Water Soluble and Fuel Soluble Biocides

WATER SOLUBLE BIOCIDE		FUEL SOLUBLE BIOCIDE	
Advantages	**Disadvantages**	**Advantages**	**Disadvantages**
Lower treat rates and treat cost	Possible concern with disposal of biocide treated water bottoms	Long lifetime in fuel; biodegradation is usually not a problem	Higher volumes of biocide are needed due to larger volume of fuel compared to water
Typically has no negative impact on fuel performance properties	Short-term effectiveness due to biodegradation of biocide	Will contact plaques which may reside on sidewalls of tank which are above the water interface	Treat cost is typically higher
Rapid kill at the interface	Frequent treatment may be necessary		High treat rates may have negative impact on certain fuel performance properties

Some common compounds which are approved for various fuel-water biocide applications include glutaraldehyde, methylene bis-thiocyanate and 2-methyl-4-isothiazoline-3-one. Characteristics of these and other common biocides are provided in **TABLE 6-3.**

TABLE 6-3. Characteristics of Common Biocides

Biocide	Bacterial Activity	Typical Dosage* (Based on Water Content of System)	Stability
Glutaraldehyde	Typically effective in 3-8 hours against all bacteria	100 ppm for a 50% solution used weekly	Stable under temperature and pH fluctuations; degraded by H_2S; incompatible with polyacrylamine
Methylene-bis-thiocyanate	May take up to 12 hours to react; not generally effective against SRBs	200 ppm of a 10% solution used weekly	Degrades slowly with increasing temperature and pH
Sodium dimethyl dithiocarbamate/ disodium ethylene-bis-dithiocarbamate	Moderate effectiveness in 3-8 hours against most bacteria	200 ppm of a 15% solution used weekly	Stable under temperature and pH fluctuations; reacts with dissolved iron to form a precipitate
Isothiazoline type	Typically effective in 4-12 hours against all bacteria	300 ppm of a 1-2% solution used weekly	Degrades slowly under temperature and pH fluctuations
Dibrominated proprionamide	Typically effective in 15 minutes to 2 hours against all bacteria	100 ppm of a 20% solution used weekly	Degrades readily in high temperature, high pH conditions
Hexahydro-1,3,5-triethyl -S- triazine	Typically effective in 3 to 8 hours against all bacteria	100 ppm of a 95% solution used weekly	At pH values <6.5, activity increases but lifetime diminishes

* NOTE - although these reported dosages are based on a weekly treat rate, lower dosages can be added at a more frequent rate.

E. WAX CRYSTAL MODIFIERS

Linear, branched and cyclic paraffins can all be present in fuel. At low temperatures, fuel filtration, pumpability and injection problems are primarily due to paraffinic wax. Refiners can contend with fuel wax through processing changes and blending. Examples include:

- Reducing the 90% distillation temperature and end point
- Use of a lower percentage of a straight run fraction or waxy distillates in a blend
- Utilize a lighter fuel component stream such as kerosene or jet fuel in a blend
- Blend asphaltic compounds at low levels into residual fuels
- Reducing the IBP and EP temperatures

All of these refining techniques can be used to some degree of effectiveness at minimizing the problems associated with the low temperature handling of fuel containing paraffins and wax.

In addition to refining techniques, compounds identified as wax crystal modifiers are available for use in contending with the effects of wax in fuels. Wax crystal modifiers, also called pour point depressants or cold flow improvers, are typically polymeric compounds which have the ability to crystallize with fuel wax as it forms. By co-crystallizing with wax, the modifiers typically affect a change in the size, shape and conformation of wax crystals. Other wax crystal modifiers function by dispersing or inhibiting the nucleation or growth of wax crystals within a fuel or oil.

Copolymers of ethylene vinylacetate are the most commonly utilized fuel wax crystal modifiers. Other compounds such as vinyl acetate-fumarate copolymers, styrene-ester copolymers, diester-alphaolefin copolymers, as well as alkyl carbamate compounds are effective wax crystal modifiers. These compounds differ in both chemical structure and in the extent of performance provided. See **FIGURES 6-7 and 6-8.**

The different chemical compounds used as wax crystal modifiers do not all provide ideal performance under every circumstance. Various tests have been designed to help differentiate the performance of one wax crystal modifier over another. For example, a modifier may be quite effective at controlling wax crystal formation to enable a fuel to flow by gravity from a storage tank to a pump. However, once past the pump, the modifier may not effectively reduce the wax crystal size and shape to allow cold fuel to flow effectively through a line filter. The result is wax accumulation on the filter media, plugging of the fuel filter and halting of fuel flow. A different wax crystal modifier or a product with wax dispersant properties may be required to permit effective fuel filtration.

See **FIGURES 6-9** through **6-11** for examples of wax crystal modifier performance in different refined fractions.

Fuel problems which can typically be solved by a wax crystal modifier
- Distillate fuel pour point temperatures can typically be reduced by 45°F to 60°F

(about 25°C to 33°C) using most wax crystal modifiers. Response is treat rate dependent.
- Distillate fuel filtration temperatures can be reduced by 10°F to 18°F (about 5°C to 10°C) by most wax crystal modifiers. Greater response is possible, but additive treating rate can be quite high. Also, fuels with low initial filtration temperatures often respond more effectively to wax crystal modifier treatment than high filtration temperature fuels.
- Low temperature distillate fuel pumpability and viscosity can be reduced below that of an untreated fuel. Below the cloud point, fuel which does not contain a wax crystal modifier is difficult to handle and pump effectively. Wax crystal modifiers help minimize pumpability problems related to yield stress and viscosity limited flow.
- The pour point and low temperature viscosity of a residual fuel or heavy fuel oil can be reduced by using a heavy fuel wax crystal modifier. Often, pour point reversion can be prevented by using the correct wax crystal modifier.

Fuel performance problems which typically CANNOT be solved by a wax crystal modifier
- Problems caused by wax which has already formed in fuel cannot be reversed unless the fuel is warmed to temperatures above the fuel cloud point. Warming will dissolve wax crystals. Upon further cooling, the modifier will then function and inhibit wax related problems such as wax deposition onto interior surfaces of cold fuel lines and plugging of fuel filters.
- Reduction of fuel viscosity at high temperatures. At temperatures above the cloud point, wax in fuel is not organized into a lattice-like network or into an organized crystalline form. Above the cloud point temperature, fuel viscosity is influenced primarily by the chemical composition and concentration of all fuel components.

FIGURE 6-7. An Ethylene Vinylacetate Copolymer

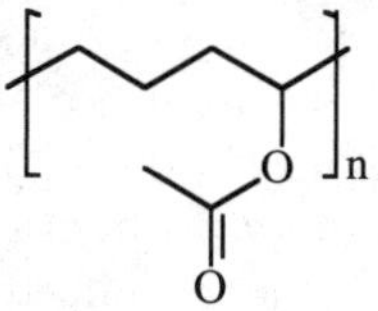

FIGURE 6-8. A Styrene/Ester Type Copolymer
(R and R'= Ester Groups)

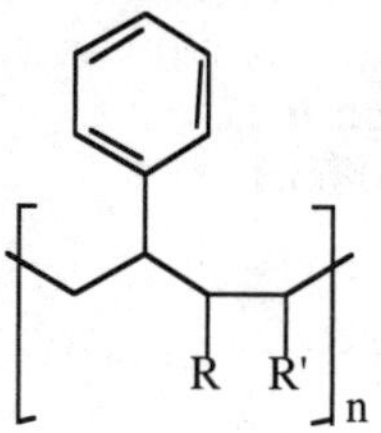

Solving Fuel Problems by Using Chemical Additives 147

FIGURE 6-9. Typical Response of an EVA in a +18°F Cloud Point Hydrotreated Low Sulfur Diesel Fuel

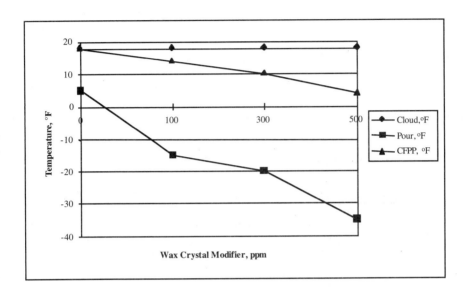

FIGURE 6-10. Typical Response of a Heavy Fuel Wax Crystal Modifier in a High Pour #6 Fuel Oil

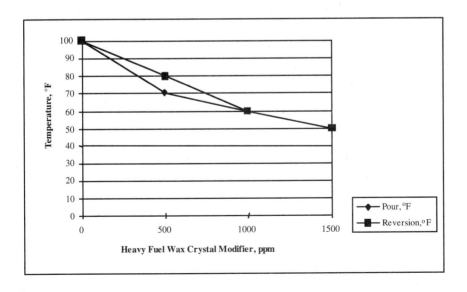

FIGURE 6-11. Pour Point Response of a EVA in a +65°F Pour Atmospheric Gas Oil

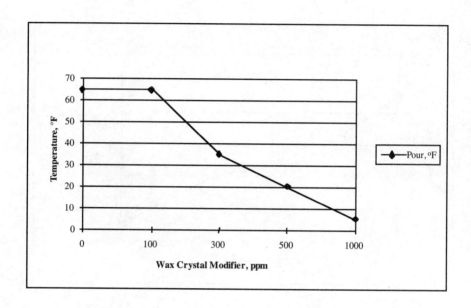

F. CORROSION INHIBITORS

In fuel systems, iron and steel corrode when both water and air enter the system. Water can contaminate fuel through refinery processes such as water washing and caustic washing, through condensation of water from the air or through other external sources. Air entrainment during pumping, mixing and blending operations can introduce the oxygen needed for corrosion to continue.

Vast amounts of information have been written about ferrous metal corrosion and corrosion mechanisms. For the purpose of this discussion, only a brief review of corrosion chemistry and corrosion process will be presented.

1. Corrosion Chemistry

The process of ferrous metal corrosion is demonstrated in **FIGURE 6-12.**

FIGURE 6-12

(A) Oxygen concentration cell corrosion beneath a deposit; (B) Oxygen concentration cell corrosion in a beaker containing an aerated piece of steel and an unaerated piece of steel. A potential develops between the aerated and unaerated steel pieces. Steel exposed to the lower dissolved oxygen concentration corrodes. Beneath a deposit (A) the oxygen-poor environment causes wastage. (From H. M. Herro and R. D. Port, *The Nalco Guide to Cooling Water System Failure Analysis,* McGraw-Hill, Inc.)

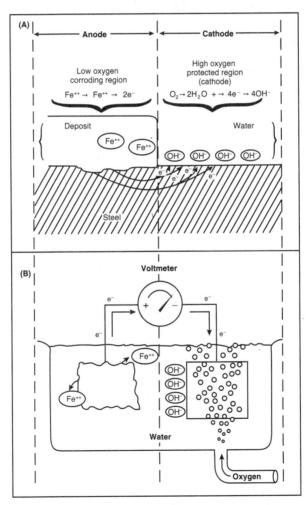

Ferrous metal corrosion can occur quite readily whenever a difference in electrical potential exists between iron and a corroding agent. Corrosion begins whenever electrons leave the iron surface at the anode and ferrous ions form:

$$Fe \rightarrow Fe^{++} + 2e^-$$

Reaction of water with the ferrous ions results in the formation of a gel-like ferrous hydroxide compound and hydrogen ions. Under acidic conditions, the hydrogen ions react with the two electrons liberated from the above reaction to produce hydrogen gas:

$$Fe^{++} + 2H_2O \rightarrow Fe(OH)_2 + 2H^+$$

$$2H^+ + 2e^- \rightarrow H_2$$

Ferrous hydroxide can continue to react with water and oxygen to form colloidal ferric hydroxide:

$$4\,Fe(OH)_2 + O_2 + 2\,H_2O \rightarrow 4\,Fe(OH)_3$$

Also, ferrous hydroxide can further react with oxygen from the air to form ferric oxide (rust) and water:

$$4Fe(OH)_2 + O_2 \rightarrow 2Fe_2O_3 + 4H_2O$$

In the formation of ferric oxide, only iron and oxygen are consumed. Water is conserved:

$$4Fe + 3O_2 + 8H_2O \rightarrow 2Fe_2O_3 + 8H_2O$$

A rusted anode appears as a raised orange to brown upper layer occupying more volume than the original iron surface. This expansion loosens the rusted layer to expose even more surface to corrosive attack.

The corrosion rate of steel increases proportionally with temperature up to about 80°C (180°F) in open air systems. At higher temperatures, oxygen is driven from the system and corrosion rates decline significantly.

The effect of dissolved oxygen on the rate of mild steel corrosion is shown in **FIGURE 6-13.**

FIGURE 6-13.

Effect of oxygen concentration on corrosion of mild steel in slowly moving water containing 165 ppm $CaCl_2$; 48-hour test, 25°C.

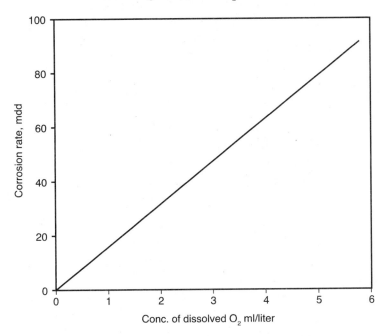

2. Hydrogen Sulfide Corrosion

Although uncommon in most fuel systems, ferrous metal corrosion initiated by hydrogen sulfide can occur. In the presence of water, hydrogen sulfide can react with iron to form iron sulfide:

$$Fe + H_2S + H_2O \rightarrow FeS + 2H^+ + H_2O$$

The iron sulfide which forms adheres to the metal surface and acts as a cathodic site. Severe pitting type corrosion can occur under these iron sulfide deposits. If oxygen is also introduced into the system, the rate of corrosion can increase dramatically.

3. Basic Forms of Corrosion

Uniform Corrosion

Metals develop a natural corrosion resistant film when exposed to the environment. Examples include the rusting of iron, tarnishing of silver and the formation of the patina on copper. These "passive" films help prevent further corrosion. However, films do not provide complete resistance to chemical attack and are destroyed by various corrosive agents.

In a corrosive medium, metals and metal alloys form reaction products which are stable and similar to those found in nature. Loss of metal in this manner is uniform over the surface and can be measured. Standard corrosion tables contain this type of information.

In general, three broad classifications of uniform corrosion based on metal loss are utilized. They are described as follows:

Classification	Metal Loss Rating
Good corrosion resistance	<5 mpy
Satisfactory if a high corrosion rate can be tolerated	5 - 50 mpy
Usually unsatisfactory	>50 mpy

Galvanic Corrosion

This type of corrosion occurs whenever two different metals are contained in a corrosive or electrically conductive medium. It can also occur when two similar metals are interconnected but are in contact with two different electrically conductive mediums.

Under either of these circumstances, the more active metal (less noble) serves as the anode and the less active metal (more noble) as the cathode. Metal loss from a small anode can be extremely high if connected to a cathode with a larger area. The closer together metals are in the galvanic series, the less the rate of corrosion. The further apart, the greater the corrosion rate. See **APPENDIX 4** for a listing of the galvanic series of metals and alloys.

Crevice Corrosion

When metal parts make contact, or when metals contact a nonmetallic part such as a gasket, the crevice formed at the point of contact can become a corrosion site. Corrosive attack at this site is usually more severe than the surrounding area due to factors such as oxygen deficiency, acidic changes or inhibitor depletion. Using nonabsorbent gasket material and avoiding threaded joints can help limit crevice corrosion.

Stress Corrosion Cracking

When sensitive metals are exposed to certain environmental factors or tensile stress, corrosion can occur. Exposure to only a few parts per million of a corrosive agent can initiate stress corrosion in some metals. High concentrations are often not necessary. Also, temperature and pH can be influencing factors.

Metal components under compression do not usually undergo stress corrosion cracking. Welding, bending, denting and installation can initiate cracking. An applied tensile stress approaching the yield strength of the metal is needed to initiate failure.

The addition of nickel as an alloying element helps to prevent stress corrosion cracking of many metals, as shown below.

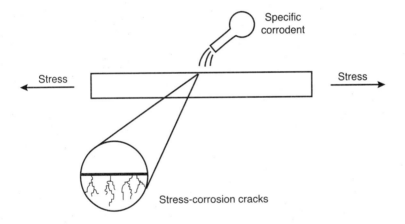

Pitting

Corrosion described as pitting is usually localized and can be in the form of small, deep pits as well as large shallow pits. It can occur in areas of stagnation where an anode site can develop.

4. Corrosion Inhibitor Function

Most fuel system corrosion inhibitors function by strongly adsorbing onto the exposed surfaces of metal to form a protective film-like layer. This inhibitor layer acts as a barrier to prevent water from contacting the surface of metal system components and initiating corrosion.

Filming inhibitors are characterized by possessing a polar, electron dense "head" and a non-polar, oil soluble "tail." Examples of compounds which have these properties include sulfonates, fatty acids and imadazolines. These materials are fuel soluble, non-volatile and quite effective at preventing water initiated corrosion of ferrous metal surfaces.

$$CH_3(CH_2)_6\underset{|}{CH}(CH_2)_7CO_2H$$
$$HO_2C(CH_2)_7CH = C(CH_2)_7CH_3$$

A Dimerized Fatty Acid Type Corrosion Inhibitor

At elevated temperatures, 450°F (232.2°C) to 500°F (260.0°C) for example, film forming inhibitors loose their effectiveness. However, during transportation and storage, finished fuels are not exposed to these high temperatures. It is only during initial refining and final combustion that fuels will be exposed to these high temperatures.

Fuel system corrosion inhibitors must have a low tendency toward emulsification with water and toward foam enhancement in turbulent systems. These properties are especially critical whenever inhibitors are used in jet fuel. The sensitivity of jet fuel pumping and injection systems requires that fuel be free of emulsions and foam.

5. Applying Corrosion Inhibitors

Corrosion inhibitors used to protect fuel system components such as storage tanks, pipelines, and combustion system equipment are typically dissolved in the fuel and delivered to the metal surface with the fuel. The inhibitor is deposited onto exposed metal surfaces as the fuel passes through the fuel distribution and handling system.

When first used in new systems which have not been treated with a film-forming inhibitor, the treat rate of inhibitor should be maintained at least twice the recommended rate for a period of about one month. This will provide enough time for an inhibitor film to develop on the exposed metal surfaces. After one month, the inhibitor rate can be reduced.

Care should be taken when using a corrosion inhibitor in an existing system which has not received corrosion inhibitor treatment in the past or in any system containing rust. When applied, a filming inhibitor will loosen and remove existing rust. The rust will travel with the fuel and accumulate on filters or in areas of low turbulence within a fuel system. New metal surfaces will be exposed upon removal of the rust and can further corrode if not protected.

Under these circumstances, the corrosion inhibitor should be added initially at low treating levels and gradually increased to the recommended level. This incremental addition will help minimize the negative effects of loosened rust or rouge and gradually build an inhibitor film on the newly exposed metal surfaces.

Also, undertreatment in this type of system is not recommended because of the continual exposure of new metal whenever rust is removed. If inhibitor is not present in adequate amounts, localized, pitting type corrosion will continue.

6. Fluid Friction Effect

The protective film formed by corrosion inhibitors can act as a lubricating film to help in reducing fluid friction within a pipeline. This property can be measured and is termed the "C Factor." This friction factor is expressed as follows:

$$C = \frac{162.04 \, QS^{0.54}}{p^{0.54} \, D^{2.63}}$$

where:
- Q = Barrels per hour
- p = Pressure drop per mile, psi
- S = Specific gravity at operating temperature
- D = Internal diameter of pipe, inches

C Factor ratings between 155 and 160 are typical for newly constructed pipelines or pipelines which have been effectively treated with a film forming corrosion inhibitor. A low C Factor indicates higher internal friction within a pipeline system, thus a reduction in efficiency.

7. Copper Corrosion

Copper can be present in fuel systems in the form of heating coils, cooling coils, brass fittings or bronze parts. Copper is quite resistant to corrosion by water but can be attacked by ammonia and sulfur compounds. Finished fuels usually do not contain ammonia unless it somehow carries over from refining process operations. Sulfur compounds such as hydrogen sulfide and possibly elemental sulfur are more frequently the cause of copper corrosion problems in fuel systems.

The anodic corrosion of copper by hydrogen sulfide yields copper sulfide and hydrogen:

$$Cu \rightarrow Cu^{++} + 2e^-$$

$$Cu^{++} + H_2S \rightarrow CuS + 2H^+$$

Under acidic conditions, hydrogen gas evolves:

$$2H^+ + 2e^- \rightarrow H_2$$

G. FUEL SWEETENING ADDITIVES

Compounds such as H_2S and mercaptans are not commonly found in finished fuels. However, the possibility of carryover of these compounds into finished fuels can occur if fuel processed from high sulfur crude oil is not properly stripped, caustic washed or sweetened through refining. If H_2S and mercaptans are found in finished fuel, they can still be removed by the addition of chemical sweetening additives.

Amine compounds including primary amines and amine-aldehyde condensation products are commonly utilized as fuel sweetening applications. Primary amines will react with H_2S to form amine sulfide compounds. These products are somewhat unstable and may tend to solubilize into water.

$$R\text{-}NH_2 + H_2S \rightarrow R\text{-}NH_3^+SH^-$$

Amine-aldehyde condensation compounds are believed to react with H_2S to form more complex carbon-sulfur bonded products.

Mercaptans are more difficult to remove from fuel than H_2S. Amine hydroxide compounds have been shown effective at removing mercaptans from fuel. The reaction products are typically not fuel soluble and must be separated before the fuel can be used.

H. CETANE IMPROVER

The most common compound used to improve the cetane number of distillate fuel is 2-ethylhexyl nitrate, an isooctyl nitrate. Dibutyl peroxide, certain glycols and other nitroparaffins are also effective at improving the cetane number of diesel fuel but their acceptance is not as widespread as isooctyl nitrate.

The Federal diesel fuel oil specification VV-F-800D lists the following compounds as acceptable cetane improvers:

a. Amyl nitrate
b. Isopropyl nitrate
c. Hexyl nitrate
d. Cyclohexyl nitrate
e. 2-Ethylhexyl nitrate
f. Octyl nitrate

Important physical properties of 2-ethylhexyl nitrate are provided in **TABLE 6-4**.

TABLE 6-4. Typical Physical Properties of 2-Ethylhexyl nitrate

Property	Typical Result
Flash Point, °F	>150
Pour Point, °F	<-50
Specific Gravity @ 60°F	0.964
Viscosity, cSt @ 60°F	1.9
Viscosity, cSt @ 32°F	2.8
IPB (decomposes on boiling), °F	279
Autoignition Temperature, °F	374

A thorough discussion of cetane improver performance and applications can be found in **Chapter 4**.

I. DETERGENTS AND DISPERSANTS

The addition of detergents and deposit control additives to fuels is a common practice today. Gasoline and diesel fuel may often contain significant levels of compounds formulated to prevent deposit formation on and within fuel intake system components.

In gasoline engines, most detergents will prevent deposit formation on carburetors, fuel injectors and intake valves. Some detergents also possess the ability to remove existing deposits from intake valves and control the formation of deposits within the combustion chamber of an engine.

Detergents used in diesel fuel help to control deposit formation on fuel injector nozzles and act to prevent corrosion of the nozzle orifices. Diesel fuel detergents also aid in preventing deposit and gum formation on high pressure fuel injector parts.

Jet fuel, burner kerosene, heating oil and heavy marine fuel oils do not typically contain detergents. The widespread need for detergents to improve fuel performance in these applications has not yet developed. Although in some small markets, combustion catalysts and burner nozzle antifoulants are utilized.

1. Deposits in Fuel Systems

a. Carburetors and Fuel Injectors

Fuel olefins have been implicated as the primary cause of deposits in gasoline fuel injectors and carburetors. High boiling, high molecular weight aromatic components have also been shown to contribute to intake system deposit and gum formation. Once formed, other compounds in the fuel can adhere to these deposits to form an amorphous-type deposit.

b. Intake Valves

Deposits on intake valves are also caused primarily by olefins, especially diolefins, and higher molecular weight aromatic fuel components. The higher temperatures within the intake valve region promote the formation of deposits which are more carbonaceous than those which form on fuel injector tips. Also, components of the lubricating oil additive package are often found within the matrix of intake valve deposits.

Intake valve deposits (IVD) vary in appearance and character. Some are hard, dry and crusty, some are relatively soft and thick, while others are thin and varnish-like. They can form on valve tulips, valve stems and on valve seats. Poor starting, warmup and initial acceleration are common symptoms of an IVD problem.

c. Combustion Chamber Deposits

Much work is being performed by engine manufacturers, oil producers, additive companies and others to identify the root cause of combustion chamber deposits. Sources of deposits which form within the combustion chamber of a gasoline engine can be fuel induced or additive induced. Components such as high molecular weight aromatics in fuel and certain fuel detergent additive formulations have been identified as sources of combustion chamber deposits. Often these compounds can be found in the unwashed gum fraction of gasoline. Also, high molecular weight aromatics from the lubricating oil, especially bright stock, are believed to contribute to combustion chamber deposit formation.

Recent investigation has shown that fuel induced deposits first form as oxidized compounds which condense onto the surface of the combustion chamber. Continued heating and polymerization results in the formation of a high molecular weight liquid-like material. Heat transforms the material into a deposit composed of a crust-like outer layer and liquid underlayer. The crust-like layer can crack and flake off exposing the liquid underlayer. New deposits then form on the underlayer.

Additive induced combustion chamber deposits form near the intake valve. The deposit forms directly onto the surface and has a waxy appearance. These deposits do not undergo alterations as do fuel derived deposits. Also, additive induced deposits do not readily flake off of the combustion chamber surface.

Also, the effect of engine design on combustion chamber deposit formation is being investigated.

2. Gasoline Engine Detergents

a. Carburetor and Fuel Injector Detergents

These products have been marketed for many years. They are effective at preventing deposit formation and corrosion of carburetors and fuel injectors, but often do not prevent deposit formation on intake valves. Most products contain an amine based detergent, demulsifier and sometimes an antioxidant and metal deactivator. The effective treat rate for preventing deposit formation on carburetors and fuel injectors can range from a minimum of 20 ppm to a maximum of about 75 ppm.

b. Intake Valve Deposit (IVD) Control Detergent

With the introduction of fuel injected vehicles during the mid-1980s, it was found that some carburetor/fuel injector detergents actually contributed to the formation of deposits on intake valves, especially when used at high concentrations. For this reason, it became necessary to determine whether fuel detergents contributed to IVD problems or helped to prevent IVD deposits.

Actual on-road tests in the U.S. using a 1.8 L, 4-cylinder, automatic transmission BMW 318i have been used to determine the ability of gasoline detergents to prevent the formation of intake valve deposits. In Europe, stationary engine tests using the Opel Kadett 1.2 L 12S and the Daimler-Benz 2.3 L M 102E engines have been commonly used to qualify IVD additives. Today, other European, Japanese and U.S. built engines are being considered for IVD testing.

c. ORI and Emission Control Additives

Compounds which control the formation of deposits with the engine combustion chamber and prevent fuel octane requirement increase (ORI) are also marketed and sold. Typically, their effectiveness is seen during the early stages of use, but effectiveness may taper-off with continual use. Work is continuing to improve the performance of these additives. Also, all organic exhaust emission control additives are in the early stages of development and testing.

In a gasoline engine, the fuel detergent typically provides the following performance:
- Prevent and remove deposits on fuel injectors and carburetors
- Prevent the formation of deposits on intake valves

In addition to the performance described, some gasoline detergent formulations provide the following:
- Removal of existing deposits from intake valves
- Control the buildup of combustion chamber deposits

3. Diesel Engine Detergents

High pressure diesel fuel injection systems contain expensive and sophisticated components. The high pressure pump and injector are the key components to ensuring proper fuel management within the diesel engine. Clearances and tolerances between moving parts of the fuel pump are quite fine. Even a small amount of deposit, contamination or corrosion can significantly alter the efficient performance of the fuel injection system.

Diesel fuel detergents are usually organic amines, imides, succinimindes, polyalkyl amines or other polyamine-type compounds. These compounds function by dispersing and solubilizing deposit forming compounds into the fuel. Some detergents can also remove existing deposits from fuel injection system parts.

Detergent treat rates can typically range from 100 to 400 ppm, but may vary depending upon the fuel type and engine test method utilized. Their detergent performance can be rated by testing in compression-ignition engines such as the Puegeot XUD-9 or the Cummins L-10.

Deposit control additives or detergents are added to diesel fuel to help prevent the formation and accumulation of deposits on fuel injection system parts. Detergents provide the following performance function in the diesel engine:

- Maintain cleanliness of the fuel pump and component parts
- Help prevent fouling and deposit formation on fuel injector nozzles
- Control acid initiated corrosion and erosion of fuel injector nozzle orifices
- Provide corrosion control and protection of ferrous metal parts

J. LUBRICITY IMPROVER

Lubricity is a term used to describe the ability of a fluid to minimize friction between, and damage to, surfaces in relative motion under load. Fuel helps to lubricate and prevent wear of high pressure fuel injection pump components, especially under boundary lubrication conditions. Boundary lubrication is defined as a condition whereby friction and wear between two surfaces in relative motion are determined by the properties of the surfaces and the bulk properties of the contacting fluid. The bulk viscosity of the fuel or oil is not a critical factor under boundary lubrication conditions.

Additives identified as lubricity improvers are sometimes used to improve the wear inhibiting properties of low viscosity #1 diesel fuel and higher viscosity hydrotreated low sulfur #2 diesel fuel. Lubricity improvers are available as single products or may be contained within a diesel fuel performance additive package. In addition to distillate fuel lubricity, gasoline lubricity is being investigated.

Certain compounds which have the ability to film or adsorb onto a metal surface are effective at improving fuel lubricity performance. These compounds include modified fatty acids, modified fatty amines and other amine based compounds. For years, the lubricity performance of jet fuel has been improved by treatment with organic acid based corrosion inhibitors.

Diesel fuel lubricity performance is presently evaluated by using actual pump rigs or by methods such as the Southwest Research Scuffing Ball-on-Cylinder Lubricity Evaluator (BOCLE), High Frequency Reciprocating Rig (HFRR) and the Ball-on-Three Disks (BOTD). The fuel industry is presently making determinations on how to place specifications on fuels evaluated by the Scuffing BOCLE and HFRR methods. The Scuffing BOCLE and HFRR methods are contrasted in **TABLE 6-5**.

TABLE 6-5. Comparison of Scuffing BOCLE and HFRR Fuel Lubricity Test Methods

Operating Parameter	Scuffing BOCLE	HFRR
Fuel Sample Size	50 ± 1.0 ml	1 ± 0.20 ml
Test Cell	A rotating test ring is partially submersed into fuel; stationary test ball with applied load is lowered onto rotating test ring	A test disk is immersed into fuel; a vibrating test ball with an applied load is lowered onto the test disk
Frequency	-	50 ± 1 Hz
Vibrational Path Length	-	1 ± 0.02 mm
Ring RPM	525 ± 1	-
Initial Applied Load	500 g	200 ± g
Test Time	30 s break-in time 60 s test time	75 ± 0.1 min
Test Fluid Temperature	25 ± 1°C	25 ± 2°C for volatile fluids 60 ± 2°C preferred temperature for most fluids
Incremental Applied Load	500 - 5000 g	-
Additional Comments	Relative humidity and air flow through the system must be controlled within specified limits. The test ring and test ball must be properly prepared before use. Improper preparation and contamination can impact test results	Relative humidity of air in the system must be greater than 30%
Result	Friction coefficient is determined at applied load. Presently being considered is a determination of applied load relative to a friction coefficient of 0.175. This determination, however, is still under consideration	Wear scar diameter on the test ball is determined. Presently being considered is a wear scar diameter of 0.44 mm or less for acceptable fuels. This determination, however, is still under consideration

K. PROBLEMS RELATED TO FUEL ADDITIVES

Although not common, fuel additives can cause problems in fuel systems. The problems often are due to the following:

- The additive has inherent physical property limitations.
- The additive may interact with a fuel contaminant such as water or rust.
- The additive may react with another chemical component in the fuel or with other fuel performance additives.

Some of the problems encountered are described as follows:

1. Antioxidants

a. Crystallization

When cooled or contaminated, hindered phenolic antioxidants and some PDA antioxidants are known to precipitate from solution as concentrated compounds. At temperatures below 60°F (15.5°C), some antioxidants will begin to crystallize and drop from their solvent carrier as relatively pure materials.

Fortunately, upon warming to temperatures of 75°F (23.9°C) to 100°F (37.8°C), these crystals will dissolve and return back into the solvent carrier. The performance of the antioxidant is usually not adversely affected by this process.

These antioxidants are also sensitive to contamination by impurities such as rust, dirt or some other material. The presence of these impurities can initiate the formation of an antioxidant seed crystal. Once a seed crystal forms within the concentrated additive, it can cause the volume of stored product to crystallize. For this reason, keeping antioxidants free of contamination helps to ensure product integrity and performance.

b. Saybolt Color Degradation

Phenylenediamine (PDA) type antioxidants are naturally very dark in color. If used in water white fuels such as kerosene, #1 diesel or even certain hydrocarbon solvents, the color can be degraded to failing Saybolt color values.

c. Discoloration of Vehicle Paint

Phenylenediamine gasoline antioxidants will discolor paint if allowed to remain on the exterior finish of a vehicle. The PDA antioxidants will react with oxygen to impart a color change to the paint. It is always recommended that gasoline which spills or overflows onto a painted surface be washed off as soon as possible.

2. Corrosion Inhibitors

a. Rouge Removal

Rouge is a term used to describe the fine, almost powder-like corrosion material which develops on the inside walls of fuel system equipment. Pipelines, storage tanks and transfer lines which have rust buildup and accumulation on the internal surfaces are continually releasing low levels of rouge or rust into fuel.

When fuel containing a corrosion inhibitor is utilized in systems containing rouge, the inhibitor "films out" on the rouge as it would on clear steel. As a result of this filming effect and inhibitor buildup, the rouge is broken free from the pipe or tank wall and released into the fuel.

As the fuel passes through various tank and line filters, rouge can be trapped by the filter media. Filter plugging and halting of fuel flow through the filter can result.

For this reason, it is recommended that corrosion inhibitor addition to previously uninhibited systems be closely monitored. Initial low level addition followed by a gradual increase in the corrosion inhibitor treat rate will help to minimize the removal of pipeline rouge and filter plugging problems.

b. Filter Plugging

When fuel processing involves washing with a caustic solution for any reason, some of the caustic may carry over into the finished fuel blend. Carboxylic acid based corrosion inhibitors are very sensitive to the presence of caustic. The acid functionality can react with caustic to form a gel-like organic salt. This material readily accumulates on filters and can eventually plug the filter.

c. Injection Pump Sticking

A fatty acid based corrosion inhibitor can form a tenacious film on the metal surfaces of a fuel injection pump. The tight tolerances in high pressure injection pumps require very little fluid film for proper lubrication.

In recent years, the concern over fuel lubricity has led to the use of high concentration of corrosion inhibitors to improve diesel fuel lubricity performance. The film formed on the metal surface by fatty acid type corrosion inhibitors improves fuel lubricity. However, the result of this overtreatment may lead to sticking of injection pump parts.

d. WSIM Degradation

The ASTM D-3948 Water Separation Index, Modified (WSIM) test is used to identify the emulsifying tendencies of additives in jet fuel. High concentration of film forming corrosion inhibitors have been shown to severely degrade the water separation tendencies of jet fuel. Treat rates as low as 20 ppm of some inhibitors can degrade the WSIM to a failing rating.

3. Demulsifiers and Dehazers

a. Precipitate Formation

Demulsifiers are quite effective at clearing fuel of water initiated haze. Occasionally, upon clearing the fuel, the water will be removed by interacting with a demulsifier to form an emulsion. This emulsion will precipitate from the fuel and settle to the bottom of a fuel storage tank. This emulsion can plug filters and clog small lines if not removed from the tank with the water bottoms.

b. Creation of Haze in Pipeline Fuel

To remove water haze from cloudy fuel, refiners will sometimes use fuel dehazers. Some dehazers disperse water into fine droplets so that the water haze cannot be seen. The fuel, thus, appears bright and clear. However, if fuel treated with a dehazer is transported through a pipeline system, the dispersed water can sometimes cling to the pipeline wall and remain as waterdroplets.

A problem can then occur with the batch of fuel which follows in the sequence of pipeline shipments. The water on the pipeline wall can be picked up by the subsequent fuel batch. The result could be the creation of haze in this subsequent fuel shipment.

c. Insolubility in Additive Packages

Gasoline and diesel detergent additive packages contain compounds to prevent deposit formation and fouling within internal combustion engines. Demulsifiers are added to prevent water from emulsifying with the detergent component. Due to high molecular weight and polarity, some demulsifiers are not freely soluble in detergent formulations. Problems in maintaining demulsifier solubility within the package can develop. To resolve this situation, cosolvents are sometimes used.

d. Dispersion of Line or Tank Deposits

If high concentrations of dehazing compounds are added to fuel, >1000 ppm for example, a possibility exists for the dehazer to interact with deposits held within a fuel transfer line or fuel tank. The surfactant-like character of the dehazer may enable it to function in removal and dispersion of existing deposits.

For this reason, extreme care should be taken when using high concentrations of dehazers to clear fuel of water haze. The excess dehazer may also act to solubilize deposits which are present within a fuel system. The deposits may then be carried with the fuel throughout the distribution system.

4. Pour Point Improvers - Wax Crystal Modifiers (WCM) - Cloud Point Improvers

a. Reversion of Pour Point and Filtration Temperatures

In order to achieve very low pour points and filtration temperatures in some fuels, high levels of wax crystal modifiers must be used. On occasion, high concentrations of a WCM can cause the pour point or filtration temperature to reverse. This is due to the influence of the high viscosity of these inhibitors at the low test temperatures.

b. Poor Solubility in Paraffinic Solvents

Many of the concentrated wax crystal modifiers which are used to lower the pour point of distillate and residual fuels are highly viscous. Because of this fact, it is often necessary to dilute the wax crystal modifier with solvent to keep it fluid at low temperatures.

Diesel fuel and kerosene alone are not effective solvents for use in solubilizing most wax crystal modifiers for low temperature application. The low KB value of diesel fuel and kerosene makes these fuels ineffective at holding the wax crystal modifier in solution for extended periods of time at low temperatures. Stratification and separation of the wax crystal modifier from solution can lead to application problems such as filter plugging and inaccurate wax crystal modifier treatment.

c. Water Emulsification and Fuel Haze Created by Cloud Point Improver

Certain cloud point improvers function by effectively inhibiting the nucleation of wax crystals. This can be accomplished by dispersion of the wax, thus interfering with nucleation. By functioning as an effective dispersant, certain cloud point improvers can help to solubilize water into fuel to give the fuel a cloudy, hazy appearance. As little as 200 ppm of a cloud point improver can create an opaque, relatively stable haze in treated distillate fuel.

d. Antagonism Between Pour Point Improver and Cloud Point Improver

Fuels treated with a cloud point improver (CPI) may required additional CPI treatment whenever a wax crystal modifier is used to reduce the pour point of the fuel. Often, the cloud point of a CPI treated fuel will increase whenever a pour point improver is used. To compensate for this phenomenon, additional cloud point improver must be added to recover the lost performance.

The dispersant property of the cloud point improver is impeded by the nucleation enhancement performance of a pour point improver. In other words, the cloud point improver functions to inhibit crystal growth while the pour point improver functions to maintain and control crystal growth. These two performance properties result in an antagonism between the cloud point improver and the pour point improver. Similar antagonism with cloud point improver performance can often be seen whenever a wax crystal modifier is used to reduce fuel CFPP.

A secondary problem may occur whenever CPI treated fuels are blended with other fuels in a fungible system. The effect of the cloud point improver may be lost due to dilution or limited performance in other fuels.

5. Distillate Fuel Stabilizers

a. Fuel Color Darkening and/or Filter Plugging

Some stabilizer formulations can function as dispersants to prevent the settling and accumulation of deposits in fuel systems. If fuel containing a stabilizer with dispersent properties is stored or transported in a system having existing deposits, the dispersent would act to break loose and suspend the deposits into the fuel. The resulting fuel would appear dark in color.

Also, some distillate fuel stabilizers contain metal chelating agents. The purpose of this is to help prevent the metal catalyzed oxidation of fuel components. However, if a distillate fuel containing a chelating agent is stored in a tank or transferred through lines which are rusted, the stabilizer may act to chelate iron. As a result of this, the fuel containing the stabilizer appears dark.

6. Cetane Improver

a. Fuel Darkening After Thermal Stability Testing

A common method used to evaluate the stability of diesel fuel is to heat the fuel at a temperature of 300°F (148.9°C) for 90 minutes. After that time period, the fuel is analyzed for color change and deposit formation.

Cetane improvers can interact with fuel at the 300°F test temperature. The products of this interaction initiate the formation of dark, insoluble, higher molecular weight compounds. The result is fuel which is much darker in color than test fuel which does not contain cetane improver. This problem does not occur in all fuel containing cetane improver.

b. High Conradson Carbon Number

Under the conditions of this test, the presence of cetane improver may give higher than normal carbon residue values.

c. Elastomer Incompatibility

Octyl nitrate cetane improvers are not compatible with a number of different elastomers used in additive storage and injection systems. The information in **TABLE 6-6** demonstrates that Teflon has the best compatibility with octyl nitrate cetane improver.

TABLE 6-6. Elastomer Compatibility with Cetane Improver

Elastomer	Weight Gain or Loss	Appearance
Natural Rubber	Gain	Enlarged; Swollen; Soft; Brittle
Polyethylene	Gain	Swollen; Slightly stained
Neoprene	Gain	Slightly swollen; Stained yellow
PVC	Gain	Enlarged; Swollen; Soft
Hypalon	Gain	Enlarged; Swollen; Soft; Brittle
Teflon	No Change	No Change
Buna-N	Gain	Enlarged; Swollen; Soft
Polypropylene	Gain	Softened
Plexiglas	Gain	Swollen; Soft; Pitted surface
Ethylene-Propylene	Gain	Enlarged; Swollen; Soft
Polyurethane	Gain	Swollen; Brittle; Stained brown
Viton	Gain	Enlarged; Swollen; Soft

CHAPTER 7

Test Methods Used to Identify and Solve Fuel Problems

In order to identify and solve a fuel problem, it is usually necessary to collect information about the physical and chemical properties of the fuel. A variety of test methods are available for this purpose. Outlined below are some of the primary testing methods which are commonly used to evaluate the composition and performance of fuels.

A. COMMON TEST METHODS USED TO DETERMINE THE OXIDATIVE STABILITY OF FUEL

Fuel degradation by oxidation is a common cause of fuel stability problems. Gasoline, jet fuel, diesel fuel and heating oil are all susceptible to oxidative degradation. The following methods can be used to help determine the oxidative stability of fuels:

1. Gasoline Stability

The gasoline specification ASTM D-4814 requires gasoline which is marketed and sold to possess a minimum level of resistance to oxidative degradation. This resistance is determined by test method ASTM D-525 and is outlined below:

a. ASTM D-525/IP 40
Oxidation Stability of Gasoline (Induction Period Method)
ASTM SUMMARY

> *The sample is oxidized in a bomb at 15 to 25°C (59 and 77°F) with oxygen at 100 psi and heated to a temperature between 98 and 102°C (208 and 216°F). The pressure is read at stated intervals or recorded continuously until a break point is reached.*

> *The time required for the sample to reach this point is the observed induction period at the temperature of the test, from which the*

induction period at 100°C (212°F) may be calculated.

The break point is defined as the point in the pressure-time curve that is preceded by a pressure drop of exactly 2 psi within 15 minutes and succeeded by a drop of not less than 2 psi in 15 minutes.

The result of this testing provides information on whether fuel components are susceptible to attack by oxygen. If fuel components do react with oxygen, then further degradation of the fuel can be expected. The final result could be gum formation, fuel discoloration and fuel system deposits.

Experience has shown that fuels with induction period times reported to be <240 minutes are unstable. These fuels tend to perform poorly in the field under long-term storage conditions.

USE: This testing method can be used to help anticipate or determine the following problems:
- Color change or darkening of fuel upon storage.
- Deposit formation on fuel system components (i.e., line varnish, valve deposits, etc.).
- Sludge formation in fuel system; filter plugging.

The ASTM D-4814 gasoline specification requiring an induction time of >240 minutes has been established for all gasoline sold in the U.S. A similar induction time is maintained in the U.K, Japan and several other countries. Also, even higher induction time periods are required in some countries.

If a gasoline does not meet this specification, antioxidants can be added to the fuel to provide an increase in induction time. Antioxidant treat rates of 5 to 50 ppm are typical. Also, the addition of a metal deactivator at a 1 to 2 ppm treat rate may improve induction time.

b. ASTM D-381/IP 131

Existent Gum in Fuels by Jet Evaporation

The ASTM D-4814 gasoline specification states that an existent gum weight be measured. Existent gum is measured by ASTM D-381. This method is summarized by ASTM as follows:

ASTM SUMMARY

A measured quantity of fuel is evaporated under controlled conditions of temperature and airflow or steam. For aviation gasoline and aircraft turbine fuel, the resulting residue is weighed and reported as milligrams per 100 ml. For motor gasoline, the residue is weighed before and after extracting with n-*heptane and the results reported as milligrams per 100 ml.*

Results of this test are reported as "unwashed gum" values or as "washed gum" values. The gums which are shown to exist in fuel can be either organic or inorganic in nature.

USE: From the *unwashed* and *washed* gum values, the D-381 values can be used to help identify the following:

Unwashed Gums
- High boiling fuel components.
- Oxidized high molecular weight organics.
- Fuel additives such as detergents and corrosion inhibitors.
- Inorganic materials such as iron and calcium salts.

Washed Gums
- Oxidized high molecular weight organics.
- Inorganic materials such as iron and calcium salts.

Although other compounds may be found in the "gum" fraction, the components listed constitute the major portion of the existent gum volume.

The ASTM D-4814 gasoline specification requiring an existent gum rating of <5 mg/100 ml of fuel has been established for all gasoline sold in the U.S. A similar existent gum rating is in effect worldwide. If gasoline does not meet this specification, the addition of antioxidants will NOT reduce the existent gum level. Typically, existent gum levels cannot be reduced by chemical additive treatment.

2. Distillate Fuel Stability

Distillate fuel can darken in color and can degrade to form sediment after exposure to air and/or heat for a period of time. Degradation is also accelerated by exposure to dissolved metal ions, especially copper. During this process, oxygen in the air reacts with fuel components to form compounds which are often dark in color and unstable. Also, condensation type reactions can occur and result in the formation of high molecular weight, insoluble organic compounds.

There are various test methods used throughout the petroleum industry to help determine the stability of distillate fuel. Most methods involve heating the fuel and exposing it to either air or an oxygen rich mixture for a specific period of time. The basic format for most stability test methods involve the following methodology:

After the test period, the color of the fuel is usually measured and compared to the color before testing. The color is normally measured by method ASTM D-1500. If the fuel color changes by more that 2.0 ASTM units, the fuel is considered to be unstable. After the fuel color has been measured, the test fuel is filtered through a standardized filter medium such as Whatman or Millipore filter pad. The filter pad is pre-weighed so that the total weight of insoluble compounds can be collected and measured.

Other tests require that only the color of the pad be measured after filtration of the fuel. Although no ASTM or SAE specification now exists to qualify the value of weighed fuel insolubles, an accepted industry standard of <2 mg of insoluble compounds per 100 ml of fuel is considered to be a low deposit forming, stable fuel. As a general guideline, weighed insolubles in the range of 2 mg to 7 mg per 100 ml of fuel would indicate moderate to poor stability. Total insolubles >7 mg per 100 ml of fuel would indicate unstable fuel.

The following is a listing and summary of common methods utilized throughout the petroleum industry to rate the stability of distillate fuel:

a. DuPont F21 Stability Test

Summary

A 50-ml sample of distillate fuel is immersed in a 300°F oil bath for 90 minutes. After removal from the bath and cooling to room temperature, the color of the fuel is measured and compared to the initial color. The fuel is then vacuum filtered through two #1 Whatman filter pads and rinsed with heptane. The color of the pad is compared to a standard color chart and rated. The color of the pad may also be measured by a light reflectance meter.

This short-term test can be utilized to quickly determine whether distillate fuel has a tendency to degrade in color and form deposits upon storage and use. Although no firm correlation has been established between the results of this testing and other longer term test methods, this method can be used to identify potentially unstable distillate fuels.

b. ASTM D-4625/IP 378
Distillate Fuel Storage Stability at 43°C (110°F)

ASTM SUMMARY

Four-hundred milliliter volumes of filtered fuel are aged by storage in borosilicate glass containers at 43°C (110°F) for periods of 0, 4, 8, 12 and 24 weeks. After aging for a selected time period, a sample is removed from storage, cooled to room temperature, and analyzed for filterable insolubles and for adherent insolubles.

This test method has been demonstrated effective for use in predicting the long term storage stability of distillate fuel. The results obtained closely approximate those obtained after fuel storage at ambient temperature. For example, fuel storage at 43°C (110°F) for 12 to 13 weeks has been shown to correlate well with one year of storage at ambient temperature.

Early work on the development of this test method was carried out by the U.S. Navy. It was found that approximately the same amount of new total insolubles formed in one week of beaker storage as formed in one month of bottle storage at 19°C (67°F).

c. ASTM D-2274/IP 388

Oxidation Stability of Distillate Fuel Oil (Accelerated Method)

ASTM SUMMARY

A 350-ml volume of filtered middle distillate fuel is aged at 95°C for 16 hours while oxygen is bubbled through the sample at a rate of 3 L/h. After aging, the sample is cooled to approximately room temperature before filtering to obtain the filterable insolubles quantity. Adherent insolubles are then removed from the oxidation cell and associated glassware with trisolvent. The trisolvent is evaporated to obtain the quantity of adherent insolubles. The sum of the filterable and adherent insolubles, expressed as milligrams per 100 ml, is reported as total insolubles.

This test method is commonly utilized throughout the world to rapidly determine the oxidative stability of distillate fuel. Although not as effective at predicting the long-term stability of distillate fuel as ASTM D-4625, this method is useful for measuring the resistance of fuel to rapid degradation by oxidation. Metal catalysts such as copper and iron are sometimes added to the fuel to further accelerate degradation, and to also test the performance of metal deactivators at preventing metal catalyzed fuel degradation.

d. Williams Brothers Pipeline Stability Test for Distillate Fuel

SUMMARY

A 50-ml sample of distillate fuel is placed into a stainless steel test bomb and sealed. The bomb is pressurized with oxygen at 100 psi and then placed into a heating block maintained at a temperature between 98°C to 102°C. The test bomb is held at this temperature for 16 hours.

After this test period, the bomb is removed from the heating block and cooled to room temperature. The color of the fuel is measured and the insoluble residue is weighed.

If the final fuel color is greater than ASTM 6.0, the fuel has failed to pass the test; if the heptane washed insoluble residue is greater than 50 mg/100 ml, the fuel has failed to pass the test.

This accelerated stability test method is preferred by some fuel refiners and marketers over other test methods used to determine the stability of distillate fuel. The more severe test conditions, relatively short test time period and similarity of this method to **ASTM D-873, Oxidation Stability of Aviation Fuels (Potential Residue Method)** contribute to its acceptance.

e. ASTM D-5304
Assessing Distillate Fuel Storage Stability by Oxygen Overpressure

ASTM SUMMARY

A 100-ml aliquot of filtered fuel is placed in a borosilicate glass container. The container is placed in a pressure vessel which has been preheated to 90°C. The pressure vessel is pressurized with oxygen to 800 kPa (absolute) (100 psig) for the duration of the test. The pressure vessel is placed in a forced air oven at 90°C for 16 hours. After aging and cooling, the total amount of fuel insoluble products is determined gravimetrically and corrected according to blank determinations.

USE: Separately or together, these different distillate fuel stability methods can be used to help identify the following:

- Tendency of fuel to darken in color during storage.
- Tendency of fuel to form sludge and deposits in storage.
- Effect of various metals on the storage stability of distillate fuel.
- Effect of blending cycle oil, gas oil or other fractions on the stability of distillate fuel.

B. TESTING THE COPPER CORROSION PROPERTIES OF FUEL

Hydrogen sulfide, mercaptans, active elemental sulfur, inorganic acids and ammonia can all attack and corrode copper. The presence of these compounds in fuel can lead to destruction of copper heating lines, cooling coils, and non-ferrous metal fittings. Also, hydrogen sulfide and mercaptans can contribute to fuel odor problems.

Although there is no ASTM specification for gasoline mercaptan or hydrogen sulfide levels, a copper corrosion specification does exist. Since mercaptans and hydrogen sulfide attack copper and copper containing alloys such as brass and bronze, it is important to minimize fuel mercaptans and hydrogen sulfide concentrations.

ASTM D-130/IP 154

Detection of Copper Corrosion from Petroleum Products by the Copper Strip Tarnish Test

ASTM SUMMARY

A polished copper strip is immersed in a given quantity of sample and heated at a temperature specified for the fuel or oil being tested. At the end of the test period, the copper strip is removed, washed and compared with the ASTM copper strip standards.

The gasoline specification ASTM D-4814 states that gasoline must pass the ASTM D-130 test with a corrosion rating of "1b" or better before it can be sold. The addition of sweetening additives to fuels which do not meet this specification can often improve the corrosion rating. Additive treat rates of 25 to 1000 ppm are common. The following test conditions are recommended for various fuels and oils:

Fuel or Oil	Temperature, °C (°F)	Time, h
Natural Gas	40 (104)	3
Diesel Fuel, Fuel Oil, Automotive Gasoline	50 (122)	3
Aviation Gasoline, Aviation Turbine Fuel	100 (212)	2
Lubricating Oil	>121 (>250)	Various

USE:
- This method can be used to determine whether fuel is corrosive toward copper and copper alloys. If corrosive, the fuel will tend to degrade rapidly in storage systems containing copper, bronze or brass components.

C. COMMON TEST METHODS USED TO DETERMINE THE FERROUS METAL CORROSION PROPERTIES OF FUEL

Although no ASTM ferrous metal corrosion specification exists now for gasoline, diesel fuel and other fuels, many refiners and marketers have adopted the National Association of Corrosion Engineers (NACE) Corrosion Standard Method TM-01-72-93 as a specification. This method has also been utilized by most product pipeline companies and is an established requirement. The NACE corrosion method is summarized as follows:

1. NACE Corrosion

A polished steel spindle is immersed into 300 milliliters of fuel maintained at 100°F. The fuel is continually stirred during the entire test period. Upon immersion for 30 minutes, 30 milliliters of distilled water are added to the fuel. After remaining in the fuel for an additional 3½ hours, the spindle is removed and the surface is rated for percent corrosion.

The NACE corrosion scale is a visual rating of surface rusting from 0% to 100%. Most refiners, product pipeline companies and marketers of fuel require a NACE surface rust rating of 5% or less. The NACE corrosion rating scale is outlined in **TABLE 7-1**.

TABLE 7-1. NACE Corrosion Test Spindle Rating Scale

% Surface Rust	NACE Rating
0	A
1-3 spots <1mm in diameter	B++
<5	B+
5-25	B
25-50	C
50-75	D
>75	E

2. ASTM D-665/IP 135
Rust-Preventing Characteristics of Inhibited Mineral Oil in the Presence of Water

ASTM SUMMARY

A mixture of 300 ml of the oil under test is stirred with 30 ml of distilled water or synthetic sea water, as required, at a temperature of 60°C (140°F) with a cylindrical steel specimen completely immersed therein. It is customary to run the specimen for 24 hours; however, the test period may, at the discretion of contracting parties, be for a shorter or longer period. The specimen is observed for signs of rusting and, if desired, degree of rusting.

Often, this method is further defined as ASTM D-665 A or ASTM D-665 B. Method "A" requires the use of distilled water. Method "B" requires the use of a synthetic sea water solution. The composition of this solution is found in **APPENDIX 5**.

Interpretation of the results of ASTM D-665 are often defined as either "PASS," no rusting on the spindle, or "FAIL," rusting observed on the spindle. However, a degree of rusting scale has been developed by ASTM to better define the rusting severity observed. The scale is described as follows:

Light Rusting - Rusting confined to not more than six spots, each of which is 1 mm or less in diameter.

Moderate Rusting - Rusting in excess of the above but confined to less than 5% of the surface of the specimen.

Severe Rusting - Rusting covering more than 5% of the surface of the specimen.

USE: These corrosion test methods can be used for the following purposes:
- To identify the possible source of a fuel system corrosion problem.
- To determine whether certain blend components such as an alkylate, oxygenate or cycle oil will alter the corrosion characteristics of a fuel.
- To rate the performance of fuel corrosion inhibitors.

D. COMMON TEST METHODS USED TO DETERMINE THE EMULSION TENDENCIES OF FUEL

Under ideal conditions, water and fuel should not emulsify. Fuel should naturally shed water and not solubilize water. However, ideal conditions do not always exist, and fuels can occasionally emulsify with water and appear cloudy or hazy.

The terms dehazing and demulsification are often used synonymously to describe the process of removing haze from fuel. Water in fuel is the primary cause of fuel haze and emulsion formation. Dehazing or demulsifying compounds have an affinity for water. By associating with water, dehazing compounds can remove water from fuel and help prevent water from emulsifying with other polar compounds in fuel.

Gasoline containing oxygenated organics such as ethanol, MTBE, ETBE and TAME will solubilize some of the water naturally present. Water of condensation or water dissolved in the oxygenate must not separate from gasoline when cooled. If gasoline-ethanol blends are exposed to large volumes of water, the ethanol may be extracted from the gasoline, and an ethanol-water phase separates from the fuel. Other oxygenates such as MTBE, ETBE and TAME will not separate from gasoline on exposure to high concentrations of water.

Low temperature cooling limits are specified for each month of the year in the various states of the U.S. Gasoline must meet these low temperature limits for water tolerance before it can be marketed and sold.

In addition to gasoline, water tolerance requirements also exist for jet fuel, diesel fuel and heating oil.

Unless required by ASTM, SAE, IP or some other organization, refiners and marketers typically establish their own criteria for rating fuel demulsibility. ASTM D-1094 is frequently used to rate the ability of a fuel to shed water. This method is summarized by ASTM as follows:

1. ASTM D-1094/IP 289
Water Reaction of Aviation Fuels

ASTM SUMMARY

A sample of the fuel is shaken, using a standardized technique, at room temperature with a phosphate buffer solution in very clean glassware. The cleanliness of the glass cylinder is tested. The change in volume of the aqueous layer and the appearance of the interface define the water reaction of the fuel.

TABLE 7-2 contains the descriptions provided by ASTM for rating the interface and fuel separation characteristics obtained by ASTM D-1094 testing.

TABLE 7-2. ASTM D-1094 Interface and Separation Rating Scale

Interface Conditions	
Rating	Appearance
1	Clear and clean
1b	Clear bubbles covering not more than an estimated 50% of the interface and no shreds, lace or film at the interface
2	Shred, lace or film, or scum at the interface

Separation	
Rating	Appearance
1	Complete absence of all emulsions and/or precipitates within either layer or upon the fuel layer
2	Same as 1, except small air bubbles or small water droplets in the fuel layer
3	Emulsion and/or precipitates within either layer or upon the fuel layer, and/or droplets in the water layer or adhering to the cylinder walls, excluding the walls above the fuel layer

In addition to using a phosphate buffer solution of pH 7, solutions of different pH values can often be evaluated. The use of pH 4 and pH 10 buffer solutions are typically requested. The use of pH 4 buffer simulates acid carryover into fuel from acid washing, acid extraction or acid catalysis. The use of pH 10 buffer simulates caustic washing or caustic carryover from fuel sweetening and neutralization processes.

2. ASTM D-1401/IP 79
Water Separability of Petroleum Oils and Synthetic Fluids

Other emulsion test methods are used to rate the ability of fuel to shed water. These methods include ASTM D-1401 and the "Waring Blender Test." The ASTM

D-1401 method is a lubricant test method, but has been adopted by the U.S. Military for rating diesel fuel demulsibility. This method is summarized by ASTM as follows:

ASTM SUMMARY

A 40-ml sample and 40 ml of distilled water are stirred for 5 minutes at 54°C in a graduated cylinder. The time required for the separation of the emulsion thus formed is recorded. If complete separation or emulsion reduction to 3 ml or less does not occur after standing for 30 minutes, the volumes of oil (or fluid), water and emulsion remaining at the time are reported.

Note: A test temperature of 82°C is recommended for oils with viscosities >90 cSt @ 40°C.

3. Waring Blender Test

The Waring Blender Test is not an ASTM standard, but is commonly used to rate fuel demulsibility. This test method can be summarized as follows:

SUMMARY

A 475-ml volume of fuel and 25 ml of distilled water are placed into a stainless steel, explosion-proof blending vessel. The fuel and water are stirred at low speed (10,000 RPM) for 4 minutes using an explosion-proof Waring blender. The mixture is then poured into 1-quart square bottles and the fuel layer and water layer are rated for clarity after 1, 3, 5, 8, 12 and 24 hours. A rating scale from 0 to 11 has been developed to describe fuel clarity and appearance.

4. ASTM D-3948
Determining Water Separation Characteristics of Aviation Turbine Fuels by Portable Separometer (MSEP Test)

Note: This test is often called the "Water Separation Index, Modified" (WSIM) test, named after the original water separation index procedure.

ASTM SUMMARY

A water/fuel sample emulsion is created in a syringe using a high speed mixer. The emulsion is then expelled from the syringe at a programmed rate through a standard fiberglass coalescer and the effluent is analyzed for uncoalesced water by light transmission measurement. The results are reported on a 0-to-100 scale to the nearest whole number. High ratings indicate that the water is easily coalesced, implying that the fuel is relatively free of surfactant materials. A test can be performed in 5 to 10 minutes.

Although water usually does not emulsify readily with jet fuel, certain compounds added to the fuel to improve performance such as antioxidants, corrosion inhibitors and metal deactivators may tend to emulsify and hold water in the jet fuel. Also, if present, naphthenic acid salts may initiate emulsification. The formation of ice crystals or stable emulsions in fuel could prevent the high pressure jet fuel pump from operating effectively. In order to predict the tendency of water to emulsify with jet fuel, the MSEP or WSIM test is utilized.

By adding 1000 ppm of water to the jet fuel and shearing at a high rate, the tendency of water to emulsify with jet fuel can be rated. Most jet fuel refiners place a minimum passing WSIM specification of 80% or greater on finished jet fuel. The minimum passing value accepted by the U.S. Military and the airline industry is a transmittance rating of 70%.

Ethyleneglycol monomethylether (EGME) and diethyleneglycol monomethylether (DEGME) are both approved as additives to help prevent ice crystal formation in jet fuel. At a maximum treat rate of 1500 ppm, these compounds have minimal effect on degrading the jet fuel MSEP rating.

5. ASTM D-4176
Free Water and Particulate Contamination in Distillate Fuel (Visual Inspection Procedures)

ASTM SUMMARY

In procedure 1, approximately 900 ml of fuel is placed into a clear glass, 1 liter jar and is examined visually for clarity. The sample is then swirled and examined for visual sediment or water drops below the vortex.

In procedure 2, approximately 900 ml of fuel is placed into a clear glass, 1 liter jar and examined visually for clarity. Fuel clarity is rated by placing a standard bar chart behind the sample and comparing its visual appearance with the standard haze rating photos. The sample is then swirled and examined for visual sediment or water drops below the vortex.

Both procedures 1 and 2 are performed immediately after sampling and at storage temperature conditions.

This ASTM method can be used to quickly determine the visual clarity of distillate fuel. A similar method, the Colonial Pipeline Haze Rating, has been used for many years by the finished product pipeline and fuel distribution industry to rate fuel clarity. A clarity rating of "2" or better when compared to the standard haze rating photo is considered a "passing" value throughout the fuel industry.

USE: Demulsibility testing can be used to help determine the following:
- Effect of water pH on the fuel-water separation characteristics.
- Effect of additive addition on fuel-water separation characteristics.
- Effect of high shear on fuel-water separation characteristics.
- Effect of the presence of rust and corrosion products on the fuel-water separation characteristics.

E. COMMON TEST METHODS USED TO DETERMINE THE LOW TEMPERATURE PERFORMANCE OF FUEL

Distillate Fuel

When distillate fuel is cooled, wax crystals begin to form in the fuel. The temperature at which these crystals are first observed is identified as the cloud point of the fuel. Upon further cooling, these crystals begin to coalesce into a lattice-like network to form a semisolid gel. The temperature at which this gelation occurs is termed the pour point.

For most distillate fuels, cloud point temperatures can range from 50°F to -10°F (10.0°C to -23.3°C) or lower. However, typical cloud point temperatures fall between 6°F and 16°F (-14.4°C to -8.9°C). Distillate blends containing a high paraffin content will often have cloud point and pour point values close together, sometimes within 5°F (2.8°C). Highly aromatic blends will usually have cloud and pour point values further apart in temperature.

Lowering the cloud point and pour point values of a distillate fuel can be accomplished by blending the fuel with a low wax content distillate stream such as a kerosene or jet fuel.

Also, additives are frequently used in conjunction with kerosene blending or as a substitute for kerosene blending to reduce the pour point of diesel fuel. Additives are not as frequently used to reduce the cloud point of diesel fuel.

Test methods ASTM D-97, ASTM D-2500 and ASTM D-4539 are used to evaluate the low temperature properties of distillate fuel. These methods and other industry accepted methods are reviewed below:

1. ASTM D-97/IP 15
Pour Point of Petroleum Products

ASTM SUMMARY

After preliminary heating, the sample is cooled at a specified rate and examined at intervals of 3°C for flow characteristics. The lowest temperature at which movement of oil is observed is recorded as the pour point.

The pour point test is used to determine the lowest temperature at which a fuel can be effectively pumped. However, the pour point value can be misleading, especially when it is used to determine the low temperature handling characteristics of residual fuel oil and other heavy fuels. Low temperature viscosity measurements are considered more reliable than pour point values for determining the flow properties and pumpability of these oils.

2. ASTM D-2500/IP 219
Cloud Point of Petroleum Products
ASTM SUMMARY

The sample is cooled at a specified rate and examined periodically. The temperature at which a cloud is first observed at the bottom of the test jar is recorded as the cloud point.

The cloud point test is one of the most commonly used methods to evaluate the low temperature characteristics of distillate fuel. The cloud point temperature identifies the point when wax begins to form into crystals large enough to become visible in the fuel. At this temperature, wax can settle from fuel, deposit onto fuel filters and interfere with the flow of fuel through small tubes and pipes. During cold weather months, distillate fuels with lower cloud point values are refined and blended to minimize the low temperature problems associated with wax.

Additional low temperature fuel filterability information is often required to confirm whether fuel will pass through fuel filters in cold environments. Two tests are commonly used to determine this performance value, ASTM D-4539 and IP 309. The method ASTM D-4539 is termed the Low Temperature Flow Test (LTFT) and is used to determine the low temperature filterability of diesel fuel.

3. IP 309
Cold Filter Plugging Point (CFPP)

Test method IP 309 is used to determine the low temperature filterability of distillate fuels including those treated with a cold flow improver. The Cold Filter Plugging Point (CFPP) is defined as "the highest temperature at which the fuel, when cooled under the prescribed conditions, will not flow through the filter or requires more than 60 seconds for 20 ml to pass through or fails to return completely to the test jar." This method is part of a series of standardized tests developed by the Institute of Petroleum. Test method IP 309 is summarized as follows:

IP SUMMARY

The fuel sample is cooled under the prescribed conditions and, at intervals of 1°C, a vacuum of 200 mm water gauge is applied to draw the fuel through fine wire mesh filter. As the fuel cools below its cloud point, increasing amounts of wax crystals will be formed. These will cause the flow rate to decrease and eventually complete plugging of the filter will occur.

The CFPP was developed in 1965/66 to predict the low temperature filterability performance of diesel fuels. This method is commonly used throughout the world to quickly determine low temperature fuel filterability characteristics. In more recent years, the LTFT and SFPP test methods have been developed to more precisely predict the low temperature performance properties of diesel fuel.

4. Simulated Filter Plugging Point (SFPP)

The Simulated Filter Plugging Point (SFPP) was developed in Europe as a supplement to the CFPP test. Work initiated in 1982 by the Coordinating European Council (CEC) demonstrated that the CFPP can become unreliable whenever the CFPP is lowered more than 10°C below the cloud point of the fuel. The SFPP was developed to better predict the operability of fuel at temperatures >10°C below the fuel cloud point.

Some of the primary differences between the CFPP and SFPP test procedures are outlined as follows:

Test Criteria	CFPP	SFPP
Cooling Rate	40°C/hr	6°C/hr
Sample Size	45 ml	55 ml
Filter Diameter	12 mm	7 mm
Filter Pore Size	45 microns	25 microns
Filtration Vacuum	2 kPa (8 in. H_2O)	16 kPa (63 in. H_2O), max.
Passing Value	20 ml fuel filtered in 60 seconds	5 ml fuel filtered in 60 seconds

The SFPP is summarized as follows:

CEC SUMMARY

A portion of fuel is cooled under the prescribed conditions and, at 5°C below the cloud point, is drawn into a pipette under a normal vacuum of 16 kPa through a standardized wire mesh filter. The procedure is repeated, as the fuel continues to follow the prescribed cooling regime, for each 1°C below the first test temperature.

The procedure continues to the first (highest) temperature at which the pipette cannot be filled within a time limit of 60 seconds.

5. ASTM D-4539 (LTFT) Low Temperature Flow Test

ASTM SUMMARY

The temperature of a series of samples of test fuel is lowered at a controlled cooling rate. Commencing at a desired test temperature and at each 1°C interval thereafter, a separate sample from the series is filtered through a 17 micron screen until a minimum LTFT pass temperature is obtained. The minimum LTFT pass temperature is the lowest temperature, expressed as a multiple of 1°C, at which a minimum of 180 ml of sample, when cooled under the prescribed conditions, can be filtered in 60 seconds or less.

Alternatively, a single sample may be cooled as described above and tested at a specified temperature to determine whether it passes or fails at that temperature.

In a CEC and in an independent study, the LTFT was shown to provide better correlation with low temperature vehicle operability than the CFPP.

6. Enjay Fluidity Test

This non-ASTM method is utilized to help predict the tendency of waxy fuel to plug small lines leading from outdoor fuel storage tanks. Fuel which flows from storage through smaller transfer lines may deposit wax on the internal surface of these lines or onto filter screens. Accumulation of wax can restrict or halt the flow of fuel from the system. This problem usually occurs slightly below the cloud point of the fuel.

Over the years, the Enjay Fluidity Test has been modified from a 16 – 24 hour test to a 2-hour test. Also, the original test required the use of 3400 to 3800 ml of fuel. The present test requires the use of only 40 ml of fuel. The procedure is summarized as follows:

> Place an aluminum disk into the cap of the bottom section of the test cylinder. Pour 40 ml of test fuel into the bottom section of the test cylinder. The fuel should be added at room temperature or at least 10°F (5.5°C) above the cloud point of the fuel.
>
> Screw the empty top section of the test cylinder onto the bottom section containing fuel. Place the unit into the test well and leave it undisturbed for two hours. The section of the test cylinder containing the fuel should sit on the bottom of the test well. The temperature of the test well can be adjusted to match the lowest ambient temperature conditions of a specific geographical region. Typically, the test temperatures range from -20°F to -40°F (-28.9°C to -40.0°C).
>
> After the two hour cold soak period, the volume of fuel in the test cylinder is measured. After measurement, the test cylinder is inverted so that the section containing fuel is on the top and the empty section rests on the bottom of the test well. Care should be taken not to touch the section of the test cylinder containing the fuel. Also, if the test cylinder must be removed from the cold well for measurement and inversion, minimize the removal time.
>
> After a settling time of one minute in the cold well, puncture the aluminum gasket in the cap of the test cylinder with a sharp object such as a pencil or large wire. Three minutes after the aluminum gasket is punctured, remove the test cylinder from the cold well and measure the volume of fuel which flowed through the capillary orifice separating the top and bottom sections of the test cylinder.
>
> A typical minimum passing value would be a recovery of 80% of the test fuel.

7. Amoco Pumpability Test for Distillate Fuel

This test was developed by Amoco Research to aid in determining the low temperature pumpability of distillate fuel. This method is summarized by Amoco as follows:

> Fuel is cooled to the test temperature through a predetermined temperature sequence and is then caused to flow through a length of copper tubing and filter screen similar to that found in full-scale fuel handling equipment. The rate at which the fuel flows and the volume of fuel delivered through the tube and screen system are noted as the criteria of cold flow performance or pumpability.

Typically, 300 ml of fuel are cooled from ambient temperature to a specific test temperature using a programmed cooling cycle. The time of cooling from ambient temperature to the required test temperature is usually about 16 hours. The lowest required test temperatures typically range from 0°F to -30°F (-17.8°C to -34.4°C).

After filtration through a 30-mesh filter at a nitrogen pressure of 15 psig, the amount of fuel filtered is recorded in 10 second intervals. After 60 seconds, the test is discontinued and the total amount of fuel filtered is recorded. Test results are rated as follows:

Pumpability Rating	Time to Pump 200 ml/sec	Volume Pumped in 60 sec
Good	1 - 30	—
Fair	30 - 60	—
Poor	—	100 - 200
Unsatisfactory	—	0 - 100

Residual Fuel

Crude oil and high boiling, high viscosity petroleum fractions such as #6 fuel oil, atmospheric tower bottoms and vacuum gas oil can contain wax which crystallizes at temperatures often above room temperature. It is not unusual for these oils to have base pour points of 100°F (37.8°C) or greater. In order to utilize these heavy oils, the pour point and viscosity of these oils must be reduced. One method which is used to accomplish this is to dilute the heavy oil with lower viscosity components such as diesel fuel or kerosene. The oil then becomes pumpable at lower temperatures.

Another common method which is used to improve the handling characteristics of heavy oils is to treat the oils with a wax crystal modifier. The process is similar to that used in diesel fuel treatment. Wax crystal modifiers for use in heavy oils are typically higher in molecular weight than those used in diesel fuel applications. The pour point method ASTM D-97 is also used to evaluate crude oil and heavy oils.

The behavior of wax in these oils can sometimes be influenced by temperature cycling and long-term storage. As the wax melts and recrystallizes through repeated cycles of heating and cooling, the pour point of the oil may be affected. If the pour point increases after exposure to continued heating and cooling cycles, a phenomenon called pour point reversion has resulted.

Pour point reversion can occur in oils treated with a pour point improver. Often, pour point reversion can be overcome by increasing the cold flow improver treat rate by about 25 - 50%.

8. British Admiralty Pour Point Test

This procedure can be utilized to determine whether heavy fuel wax crystal modifiers will lose their performance properties after long-term storage at fluctuating temperatures. Daily heating and overnight cooling may interfere with the ability of some wax crystal modifiers to maintain their performance properties in some residual oils and crude oils. This loss of performance is frequently termed *"pour point reversion."* The British Admiralty Pour Point Test can be utilized to help predict these reversion tendencies.

TEST SUMMARY

1) An oil sample treated with a wax crystal modifier is heated to 93°C (200°F) and poured immediately into pour point test jars fitted with a thermometer capable of reading temperatures from -37°C (-35°F) to 100°C (210°F).

2) Allow the sample in the pour point test jars to cool to 49°C (120°F), and then place the tubes into a -34°C (-30°F) cold well. Allow the test jars to remain in the bath until the temperature of the oil is -18°C (0°F).

3) Remove the test jars and allow them to warm slightly on the benchtop. Place in an oven or bath maintained at 96°C (205°F). When oil has reached 93°C (200°F), hold for an additional 30 minutes.

4) Remove the test sample and allow it to cool on the benchtop to 32°C (95°F).

5) Once cooled, place the oil sample into the pour point test well maintained at -1°C (30°F). Conduct the remainder of the test as described in ASTM D-97 by checking the pour point every 3°C (5°F).

6) Compare the pour point obtained with the value determined from a standard ASTM D-97 pour point test. If the pour point temperature obtained is higher than the standard ASTM D-97 value, a reversion in pour point has occurred.

NOTE: In addition to the 96°C (205°F) oven or bath used in Step # 3, additional ovens can be heated to 85°C, 68°C, 57°C, 49°C and 41°C (185°F, 155°F, 135°F, 120°F and 105°F). A total of six separate test samples could be evaluated for reheating temperature effect.

9. Shell - Amsterdam Pour Point Test

This method describes a procedure for determining the critical pour point of residual fuel oils.

TEST SUMMARY

While stirring, heat a sample of oil to be tested to 104°C (219°F). Once heated, pour a sample of oil into a pour point test jar to a height of approximately 57 mm. Place a cork carrying the thermometer on the test jar and set the sample into a -40°C (-40°F) test well.

When the oil reaches a temperature of -20°C (-4°F), remove the test jar from the well. If the oil is not solid at -20°C, continue cooling the test sample until it becomes solid and then remove the sample. Place the test jar in a water bath held at a temperature between 47°C and 47.5°C. When the oil sample has reached a temperature of 46°C, remove the sample and perform the standard ASTM D-97 pour point.

F. FUEL CETANE ENGINE NUMBER TESTING

In a diesel engine, air is compressed to pressures of 1000 psi or greater. Upon compression, the air temperature increases to temperatures in the range of 600 to 800°F. At close to maximum compression, diesel fuel is injected into this hot, compressed air. Ignition of fuel within this environment is initiated whenever fuel components with the lowest autoignition temperature begin to combust.

Occasionally, cetane improvers are added to distillate fuel and are capable of increasing the cetane number of diesel fuel from 1 to 14 numbers. The reason for this range of response is due to the differences in the chemical nature of the fuel. Fuel containing a higher level of aromatic compounds will typically respond less effectively to cetane improver treatment than fuels with fewer aromatic compounds. Fuel cetane number determination is evaluated by a standardized engine method ASTM D-613.

1. ASTM D-613/IP 41
Ignition Quality of Diesel Fuels by the Cetane Method

ASTM SUMMARY

The cetane number of a diesel fuel is determined by comparing its ignition quality with those for blends of reference fuels of known cetane numbers under standard operating conditions. This is done by varying the compression ratio for the sample and each reference fuel to obtain a fixed delay period, that is, the time interval between the start of injection and ignition. When the compression ratio for the sample is bracketed between those for two reference fuel blends differing by more than five cetane numbers, the rating of the sample is calculated by interpolation.

G. ADDITIONAL ANALYTICAL TESTS USED TO SOLVE FUEL PROBLEMS

There are a few additional analytical tests which are frequently used to help identify the cause of a fuel performance problem. Some of these tests are listed below:

1. ASTM D-3227/IP 342
Mercaptan Sulfur in Gasoline, Kerosene, Aviation Turbine, and Distillate Fuels (Potentiometric Method)

USE:
- Identifying the cause of copper, bronze or brass corrosion.
- Identifying the cause of elastomer degradation.
- Identifying the source of a fuel odor problem.

2. ASTM D-1319/IP 95
Hydrocarbon Types in Liquid Petroleum Products by Fluorescent Indicator Adsorption

USE:
- Identifying the paraffin, olefin, naphthene and aromatic content of fuel; good for gasoline; not as accurate for jet, diesel and other mid-distillates.
- Identifying the source of a fuel color degradation problem.
- Identifying the source of a fuel gum, sludge or deposit.
- Identify the reason for low fuel cetane number; octane number questions.
- Address fuel performance problems such as smoke, soot and power.

3. ASTM D-3120
Trace Quantities of Sulfur in Light Liquid Petroleum Hydrocarbons by Oxidative Microcoulometry

USE:
- Identifying the sulfur content of most fuels and oils.

4. ASTM D-3703
Peroxide Number of Aviation Turbine Fuel

USE:
- Predict fuel stability or instability vs. time.
- Identify the cause of elastomer degradation.

5. ASTM D-873/IP 138
Oxidation Stability of Aviation Fuels (Potential Residue Method)

USE:
- Predict the gum or deposit forming tendencies of gasoline, jet or diesel.
- Confirm complaints of fuel instability.
- Evaluate the performance of fuel antioxidants and stabilizers.

6. ASTM D-482/IP 94
Ash from Petroleum Products

USE:
- Identify whether fuel has been contaminated with lubricating oil.
- Determine the level of catalyst fine contamination in residual fuel.
- Address questions or concerns about organometallics in fuel.

7. ASTM D-86/IP 123
Distillation of Petroleum Products

USE:
- Determine whether fuels are within specification (IBP, 90%, residue, etc.).
- Identify possible contamination with low boiling or high boiling components.
- Address questions about poor fuel performance such as startability problems, slow warmup time, poor fuel economy and power.
- Identify a possible source for high deposit levels in engines.

CHAPTER 8

Identifying and Solving Specific Fuel Problems

Described below are specific problems related to fuel handling, performance and use. These problems are taken from actual field applications. Some of the problems described have more than one potential cause. By reviewing all potential causes given for each problem, it may then be possible to develop a reasonable cause-and-effect scenario using known information. If more information is needed, the steps suggested in the "What to Do" section can be taken.

Once the cause has been confirmed, one or more of the suggested "Potential Solutions" can be taken to resolve the specific problem.

PROBLEM: DIFFICULTY IN PUMPING DIESEL FUEL AT LOW TEMPERATURES / FUEL FILTER PLUGGING

Potential Causes

Wax

Measuring the cloud point, pour point, CFPP or LTFT will help to confirm whether pumpability problems are due to wax. Fuels at their cloud point, CFPP or LTFT temperatures will eventually plug fuel filters when pumped. When a fuel is very near its pour point, pumping will be quite difficult.

What to Do:
- Determine ASTM D-97, D-1500, D-4593 or IP 309 temperatures.
- Look for higher than expected filtration or pumpability temperatures.
- Look for wax on fuel filters, in-lines, etc.

Potential Solutions:
- Blend fuel with kerosene or #1 distillate to dilute the effect of wax.
- Treat with wax crystal modifier to control the growth of wax crystals.
- Use fuel with lower IBP/EP.
- Heat trace system components.

- Add solvents such as propyl alcohol or diethyleneglycol to disperse wax.
- As a last resort, add gasoline in low concentrations to dissolve wax.

Ice

Water contamination in diesel fuel is common. When diesel fuel cools, ice crystals may form in the fuel well before the fuel reaches its pour point. These ice crystals will settle to the bottom of fuel tanks and may result in fuel filter plugging and pumpability problems.

What to Do:
- Check water by Karl Fisher.
- Hold fuel at +10°F to +20°F overnight in centrifuge tube; look for ice crystals in bottom of tube.

Potential Solutions:
- Drain or filter water from fuel system routinely.
- Use an anti-icing agent or solvent in fuel such as a propyl alcohol or diethyleneglycol.

Viscosity Limited, High IBP/EP Fuel

The distillation profile will provide some indication of the low temperature performance of diesel fuel. IBP values >350°F and EP temperatures >700°F indicate that the fuel viscosity will probably be high at low temperatures. Even fuel treated with a wax crystal modifier may be difficult to pump due to the high viscosity of the fuel components.

What to Do:
- ASTM D-86 distillation; look for high IBP and EP values.
- If possible, measure viscosity vs. temperature between cloud and pour points; look for dramatic increase rather than gradual increase in slope.

Potential Solutions:
- Blend with low viscosity components such as kerosene or #1 diesel.
- Utilize lower IBP/EP fuel.

Narrow Distribution of Fuel Paraffins

Often, these fuels respond poorly to treatment with a wax crystal modifier or may require high rates of a wax crystal modifier to improve filtration and pumpability performance. This problem seems to occur whenever two situations exist:

1) the *n*-paraffin carbon number distribution for a fuel is narrow.
2) few low molecular weight *n*-paraffins are present in the fuel.

An example is provided below:

Responsive Fuel	**Poorly Responsive Fuel**
>1% *n*-paraffins @ C_7	No *n*-paraffins @ C_7
4% *n*-paraffins <C_8	<1% paraffins <C_8
Highest % of *n*-paraffins @ C_{11}	Highest % *n*-paraffins @ C_{15}
5% *n*-paraffins @ C_{20}	7% *n*-paraffins @ C_{20}

This situation can sometimes be manifested in the fuel distillation profile. More responsive fuels may have lower IBP and EP temperatures than poorly responsive fuels. Also, in poorly responsive fuels, a larger volume of fuel may distill within mid-boiling temperature fractions of the fuel distillation profile. The wax crystals which may form from these narrow boiling fractions tend to crystallize rapidly and at about the same temperature. Diesel fuel pumpability problems can be due to this narrow boiling range wax fraction.

What to Do:
- Characterize the carbon number, distribution and concentration of fuel *n*-paraffins.
- ASTM D-86 distillation and look for narrow boiling fractions.

Potential Solutions:
- Blend with low viscosity kerosene or #1 fuel.
- Treat with a wax crystal modifier effective at modifying high carbon number or narrow boiling temperature range wax.

PROBLEM: INCREASE IN POUR POINT OF RESIDUAL FUEL OIL OR CRUDE OIL AFTER HEATING OR SHEARING

An unusual behavior can be seen in some crude oils and residual fuel oils after they have been heated or have been sheared through a high RPM pump. When sampled at the pump outlet, some oils can experience an increase in pour point above that of the same oil prior to heating or shearing by pumping.

This phenomenon is not fully understood, but may be due to the effect of radiant heating and frictional heating on the redistribution and reformation of all wax into a loosely knit wax crystal lattice. The new wax lattice yields a oil with a higher pour point.

However, this new lattice structure is loosely interconnected and can readily be destroyed by continued shear. With sufficient shear, the oil again begins to flow.

What to Do:
- Heat a 100-200 ml sample of the crude oil or residual oil to 140°F or a temperature high enough to melt all wax.
- Do not heat a similar volume of a second sample of the same oil.
- Determine ASTM D-97 pour point of each sample.
- Check for increase in pour point of heated over unheated sample.

Potential Solutions:
- Utilize low temperature viscosity determinations rather than pour point values to establish low temperature handling limits on oils possessing this characteristic phenomenon.
- Low shear viscosity measurements for both heated and unheated oils should be similar.

PROBLEM: REVERSION AND ACTUAL INCREASE IN THE POUR POINT OF A CRUDE OIL OR RESIDUAL FUEL OIL

This problem differs from the problem of pour point increase after shearing and heating. In the case of pour point reversion, an increase in the pour point of a crude oil or heavy fuel oil occurs upon long-term storage.

Identifying Possible Causes

"Masking" of Wax Crystal Modifier

This phenomenon may be due to the influence of certain types of high molecular weight waxes, resins or asphaltene compounds in the oil on the performance of the wax crystal modifier. Upon initial treatment, wax crystal modifiers and asphaltenes inhibit the formation of a wax lattice network throughout the oil matrix. However, upon continued storage, the wax crystal formation becomes more organized throughout the oil. The asphaltenes and the polymeric wax crystal modifier may not be able to overcome the formation of these higher molecular weight wax crystals. The pour point of the oil, thus, increases.

Upon storage, it is also possible that the asphaltenes may interact with the wax crystal modifier to "mask" the performance of the modifier and inhibit the ability of the modifier to fully disperse in the oil. Wax crystal modifier performance may be minimized.

What to Do:
- Identify whether the oil treated with a wax crystal modifier will revert in pour point by testing utilizing either the British Admiralty Pour Point Reversion Test or the Shell Amsterdam Reversion Test.

Potential Solutions:
- If testing indicates that the oil has reversion tendencies, the addition of higher treating rates of a wax crystal modifier can sometimes overcome the reversion problem.
- Sometimes, reversion tendencies cannot be overcome by the addition of higher treat rates of a wax crystal modifier. Under these circumstances, only dilution by low viscosity products or the constant addition of heat will keep the oil fluid at temperatures below its base pour point.

PROBLEM: POOR COMBUSTION QUALITY / FUEL ECONOMY / POWER OF DIESEL FUEL

Identifying Possible Causes

High Diesel Fuel Viscosity
Highly viscous diesel fuel will not atomize properly when sprayed from a fuel injector. It can spray as large droplets or as a stream instead of as an atomized mist. Incomplete combustion of fuel can result.

What to Do:
- ASTM D-445; look for viscosity >4.0 cSt.
- ASTM D-86; look for high EP or residual material. The diesel fuel could be contaminated with high boiling, high viscosity material.

Potential Solutions:
- Add lower viscosity fuel components such as kerosene or #1 diesel fuel.
- Distill to reduce the fuel T-90 or end point temperatures.

Low Diesel Fuel Viscosity
Low viscosity diesel fuels will not penetrate and mix well into pressurized air. As a result of this, fuel will not become involved in the entire combustion chamber volume. Thus exposure to available air within the chamber is limited. Incomplete utilization of fuel volume and low power result.

What to Do:
- ASTM D-445; look for viscosity <2.0 cSt.
- ASTM D-86; look for low IBP or T-5.
- Look for contamination with high levels of low boiling, low viscosity components.

Potential Solutions:
- Distill to remove more volatile, low boiling, low viscosity materials.
- Increasing fuel viscosity by the addition of more viscous fuel components will probably not be an effective solution to this problem. The addition of viscous, high boiling components to fuel may lead to combustion smoke and soot formation.

Low Calorific Value or BTU/Gallon
The heat of combustion value for diesel fuel has a direct relationship upon the potential energy a fuel can provide. As a general rule, low specific gravity fuel has a lower BTU/gallon rating than high specific gravity fuel. This means that low specific gravity fuel will provide less available energy per gallon of fuel burned.

What to Do:
- ASTM D-86 and ASTM D-1289; determine heating value from nomograph in **APPENDIX 1**.
- ASTM D-86 T-50 temperature; if <450°F, then fuel has low heat of combustion.

Potential Solutions:
- Utilize less kerosene or #1 diesel in the fuel blend.
- Blend with higher specific gravity fuel components such as high aromatic content heating oil or stabilized FCC cycle oil. These components will improve fuel BTU ratings.

PROBLEM: POOR FLAME QUALITY OF KEROSENE
Identify Possible Causes

High Viscosity
If the viscosity of kerosene is too high, it will not readily travel up the wick of burners and kerosene lamps. Fuel consumption will be low and flame will not be bright.

What to Do:
- ASTM D-445; if >2 cSt, feed to wick will be poor.

Potential Solution:
- Dilute with lower viscosity kerosene having a higher paraffin/lower aromatic content.

High IBP
If the IBP is high, the kerosene will not be volatile enough to provide good flame height and illuminating quality.

What to Do:
- ASTM D-86; if IBP is >340°F, flame height will be poor.

Potential Solution:
- Blend with 300°F IBP kerosene.

PROBLEM: DIESEL FUEL DARKENS IN COLOR- SEDIMENT FORMS
Identifying Possible Causes

Fuel Contains Cycle Oil
Both light cycle oil and heavy cycle oil can be blended into diesel fuel. Since cycle oil contains olefinic compounds, the potential for fuel oxidation and darkening is high.

What to Do:
- ASTM D-1319 for olefins; if >2%, cycle oil may be present in fuel blend.
- If possible, check fuel blend for LCO or HCO.

Potential Solutions:
- Add a distillate fuel color degradation control additive.
- Acid wash fuel to remove olefins and color bodies.
- Clay filter the fuel to remove color bodies.
- Hydrotreat to reduce olefins and remove color bodies.

Fuel Refined from Crude Containing Polar Aromatic and Resinous Compounds
It has been determined that certain nitrogen, oxygen and sulfur containing aromatic compounds contained in crude oil can lead to darkening of refined fuels and oils. Often, these compounds are found in fuels refined from naphthenic crude oil or asphaltic, high sulfur crudes. Compounds such as indoles, quinolines and naphthenobenzothiophenes can lead to darkening of fuel.

What to Do:
- Evaluate fuel for nitrogen; if >50 ppm, darkening could be originating from organic nitrogen compounds.
- Determine whether crude source is naphthenic.
- A GC-MS analysis of the fuel, FCC fed or crude could be performed to identify the presence of polar organics.

Potential Solutions:
- Add a distillate fuel color degradation control additive at high rates before fuel enters tankage.
- Acid wash fuel to remove organic bases and other color bodies.
- Clay filter the fuel to remove nitrogen heterocycles.
- Hydrotreat to reduce organically bound nitrogen, sulfur and oxygen to ammonia, hydrogen sulfide and water, respectively.

Copper Catalyzed Oxidation of Fuel Components
Copper ions are known to catalyze the oxidation of fuel components to deposit forming compounds. Copper levels as low as 1 to 2 ppm are effective at facilitating this catalysis. Copper can originate from copper tubing as well as bronze or brass valves and fittings.

What to Do:
- ICAP for copper; look for 1-2 ppm copper.

Potential Solution:
- Treat fuel with a copper chelating agent or a metal deactivator.

Polycyclic Aromatic Hydrocarbons in Fuel
Fuel refiners will sometimes blend low levels of coker gas oil or vacuum gas oil into diesel fuel. These high boiling fractions may contain high molecular weight polycyclic aromatic compounds which can eventually precipitate to form fuel insoluble sludge and deposits.

What to Do:
- ASTM D-86; look for high residual percent - >1%.

Potential Solution:
- Some distillate fuel stabilizers possess the ability to perform as dispersants. By dispersing higher molecular weight compounds into fuel, their eventual deposition as fuel insoluble compounds can be minimized.

Naphthenic Acids in Fuel
These cycloparaffinic acidic compounds are known to form color bodies and insoluble, high molecular weight compounds in fuel.

What to Do:
- Look for the presence of these compounds by ASTM D-664-A or through specific analytical techniques used to identify cyclopentane/cyclohexane acids.

Potential Solution:
- Caustic wash fuel to remove naphthenic acids. Naphthenic acid salts are water soluble. Care should be taken to ensure that emulsification does not result.

PROBLEM: RUSTING IS IDENTIFIED ON METAL COMPONENTS
Identifying Possible Causes

Fuel Contains Solubilized Inorganic Acids
During the processing of fuels, acids such as sulfuric acid and hydrofluoric acid are used. Sulfuric acid can be used to polymerize fuel olefins and remove components from fuel such as mercaptans and thiophenes. Sulfuric and hydrofluoric acids are used in the alkylation process to produce high octane, branched paraffins. Carryover of these acids into fuel can initiate ferrous metal corrosion.

What to Do:
- Wash a sample of fuel with distilled water. Check for a reduction in pH of the wash water.

Potential Solution:
- Extract acid from the fuel with a mild caustic solution. Check pH of extracted water to ensure that the acid has been removed.

Fuel Contains High Concentrations of Water
Water contamination of fuel occurs. Water can originate from fuel processing, atmospheric condensation or external sources. Water may contain dissolved salts, may be acidic or basic, or may contain solubilized organic compounds. Water initiated corrosion can result.

What to do:
- Check fuel storage and distribution system for water such as salt water or condensate water.

Potential Solutions:
- Maintain a frequent schedule of draining accumulated water bottoms from the bottom of the fuel storage tank.
- Add a ferrous metal corrosion inhibitor to the fuel at a maintenance dosage of 5-10 ppm.

Microbial Influenced Corrosion
Microorganisms can establish growth colonies or plaques on metal surfaces. Corrosion of underlying metal can be severe. Both bacterial and fungal species can initiate this type of corrosion.

What to Do:
- Check tank bottoms and water for the presence of microorganisms.

Potential Solution:
- Treat immediately with a water soluble biocide. Continue treatment weekly to prevent further microbiological growth.

PROBLEM: FERROUS METAL CORROSION INHIBITOR FAILS TO PREVENT RUSTING OF METAL COMPONENTS

Identifying Possible Causes

Caustic Neutralization of Corrosion Inhibitor
During the processing of fuels and fuel components, solutions of sodium hydroxide or potassium hydroxide may be used. These caustic solutions can react with phenols, mercaptans and naphthenic acids to form water soluble salts. As a result, these undesirable components are removed from the fuel as the caustic separates from the fuel. Further water washing can be performed to clear the fuel of residual caustic and salts.

It is possible, however, for some residual caustic to "carry over" into the finished fuel. Residual caustic present may react with a carboxylic acid based corrosion inhibitor. If this occurs, the ability of the inhibitor to form a protective film on ferrous metal surfaces will be destroyed.

Caustic resistant corrosion inhibitors have been developed to help overcome this neutralization concern. However, these products must often be used at higher concentrations than carboxylic acid based inhibitors. Also, some caustic resistant inhibitors tend to emulsify with water later in the fuel distribution channel.

What to Do:
- Check the water in the bottom of the fuel storage vessel for the presence of high levels of potassium or sodium ions by ICAP or AA.
- Check for the presence of potassium or sodium salts of carboxylic acids by IR or FTIR.
- Check pH of the water. If >7, caustic contamination may have occurred.

Potential Solutions:
- Drain water bottoms from storage tank to eliminate contaminated water.

- Utilize a caustic resistant corrosion inhibitor. However, check for the tendency of the inhibitor to initiate fuel emulsification by ASTM D-1094. If interface emulsion is 2 or greater, do not use the inhibitor.

Acid Initiated Corrosion

Sulfuric acid and hydrofluoric acid are used as catalysts in the production of gasoline alkylate. After processing, this acid must be removed from the finished alkylate. This is typically accomplished by water washing or caustic washing the alkylate. However, if residual sulfuric acid or hydrofluoric acid remains in the fuel or alkylate, the acid can initiate corrosion. The acid is very aggressive toward initiating corrosion of ferrous metal. It is difficult for filming type corrosion inhibitors to overcome acid attack of metal.

Sulfuric acid washing is an old but effective process which can be used to remove olefins and mercaptans from fuel. Although sulfuric acid washing has been replaced by hydrotreating in many refineries, it may still be a useful process in some applications. Water washing and caustic washing are also used to remove residual acid from the fuel. Again, if the washing process is not performed or is incomplete, acid carryover is possible.

Also, caustic carryover from acid neutralization can occur, but is remote.

What to Do:
- Check pH of water bottoms. If <7.0, acid carryover and contamination may have occurred.

Potential Solutions:
- Drain contaminated water from the bottom of the storage vessel.
- Although the pH of acid water can be increased to neutral by the addition of a basic compound, this practice could result in reaction of the basic compounds with other fuel components and additives. This practice is not recommended.

Removal of Inhibitor Film

A possible but unlikely cause of corrosion would be inhibitor film removal. This can result if fuel contains high concentrations of caustic carryover from caustic washing processes.

What to Do:
- Check for the presence of caustic in the fuel by sodium analysis or by pH determination.

Potential Solution:
- Check all incoming fuel shipments for the presence of caustic in the tank bottoms. If present, drain bottoms from storage and add a corrosion inhibitor.

Reaction Between an Amine Based Filmer or Neutralizer and the Carboxylic Acid Based Corrosion Inhibitor
This problem is not common and is not expected. However, in a situation of gross overtreatment with an amine-based processing aid, interaction with a carboxylic acid based corrosion inhibitor is possible. In dilute solution within the fuel, it is unlikely that these compounds will react since only low levels, <5 ppm, of filmer or neutralizer will typically carry over into finished fuel. If the two do react, the product formed will provide corrosion protection, but not as effectively as the unreacted carboxylic acid based inhibitor.

What to Do:
- Evaluate the finished fuel for possible carryover of an amine based processing additive into the final fuel blend. Many of the analytical techniques needed to identify the presence of the inhibitor are proprietary. However, non-volatile residue FTIR and LC analytical techniques can be utilized to detect their presence.

Potential Solution:
- Utilization of a lower concentration or a different processing filmer or neutralizer may eliminate the carryover problem.

PROBLEM: HAZE OR EMULSION IS FOUND IN FUEL

Emulsification of Polar Organic Compounds
Under certain conditions, fuel detergents, corrosion inhibitors and other performance enhancing additives can emulsify with water to form water-in-oil emulsions. Conditions of low pH, high pH or high additive concentration may enhance the emulsification of polar organic compounds into fuel.

What to Do:
- Check water for low/high pH.
- Determine whether fuel is treated with performance enhancing additives.

Potential Solutions:
- Utilize an emulsion breaking additive to clear the fuel of emulsion.
- Frequently drain the accumulated water from the bottom of fuel storage system. This preventive measure alone may eliminate the possibility of water-fuel emulsification.

Presence of Microorganisms
Microorganisms living at the oil-water interface in fuel storage systems can accumulate to form a distinct "rag-like" layer of growth. This "rag" can break free and appear as an emulsion in fuel. The products of microbial metabolism are highly polar and can help to stabilize fuel emulsions.

What to Do:
- Check fuel-water interface for microbial growth.

Potential Solutions:
- Treat water with a water based biocide which is effective at killing a broad range of microorganisms. Maintain treatment weekly.
- Frequently drain water bottoms from fuel storage systems.
- The continuous application of an oil soluble fuel biocide can often help to minimize the microbial growth throughout the fuel distribution and storage system.

Contamination with Lubricating Oil
Most lubricating oils are higher in viscosity than fuels. Also, many lubricating oil formulations contain additives which provide some level of performance to the oil. If fuel becomes contaminated with lubricating oil, the possibility of emulsifying water into fuel is greatly enhanced.

What to Do:
- ASTM D-86 distillation; check residual %.
- Check ASTM D-445 viscosity.
- Check fuel and emulsion for dissolved metals.

Potential Solutions:
- Treat fuel with an emulsion breaking additive.
- Filter or separate fuel from emulsion.
- Mild heating will sometimes break emulsions.

Fuel May Contain a Cloud Point Improver
Certain cloud point improvers will disperse water into fuel and create an emulsion or haze. This phenomenon can occur at cloud point improver treat rates as low as 200 ppm.

What to Do:
- Determine whether a cloud point improver was utilized in the fuel by checking blending records.

Potential Solution:
- The addition of a fuel dehazer may provide some performance at clearing the fuel of water haze. If not, the fuel may require filtering or reprocessing to eliminate the haze.

PROBLEM: JET FUEL FAILS "WSIM" TEST
Identifying Possible Causes

Fuel Contains High Concentrations of Additives
Additives used to improve the performance of jet fuel can cause emulsification problems if added at concentrations greater than the recommended rate. Corrosion inhibitors are especially prone to emulsification if added at rates higher than their maximum effective concentration into jet fuel.

What to Do:
- Proprietary methods exist for determining the concentration of various additives in jet fuel. If available, utilize these methods to check for additive concentration.

Potential Solution:
- Clay filtration can be utilized to effectively remove polar compounds such as corrosion inhibitors, electrical conductivity improvers and antioxidants from jet fuel. Also, more costly molecular sieve techniques can be utilized to remove contaminants from jet fuel.

Fuel Is Contaminated With High Boiling, High Viscosity Components
Contamination with diesel fuel, lubricating oil or residual fuel will lower the WSIM of jet fuel. High viscosity components do not filter and shed water rapidly.

What to Do:
- ASTM D-86 distillation; check T-90 and residual for high values.
- ASTM D-445 viscosity; look for high cSt values.

Potential Solutions:
- This problem is difficult to resolve through finishing processes. Contamination with higher boiling materials may also influence the JFTOT and smoke point values. Often, fuel contaminated with other materials must be reprocessed.
- Clay filtration and molecular sieve filtration may remove enough of the higher boiling polar compounds from the jet fuel to improve WSIM.

PROBLEM: JET FUEL FAILS "JFTOT"
Identifying Possible Causes

Fuel Contains Olefins/Diolefins
Under the high temperature conditions of the JFTOT procedure, olefinic compounds could undergo polymerization-type reactions to form high molecular weight materials. These heavier compounds can foul and deposit onto the rating tube or filter screen. Tube darkening and/or filter screen plugging can result.

What to Do:
- Perform ASTM D-1319, GC-MS or other analysis to determine whether the fuel contains olefins.

Potential Solution:
- Hydroprocess to reduce fuel olefin content.

Fuel Contains Heavy Aromatic Compounds
Jet fuel specifications are written to limit the amount of aromatic compounds as well as naphthalenes. However it is possible that JFTOT failure can still occur if the aromatics present contain reactive side chains. Cycloparaffinic side groups or groups containing double bonds may react under the JFTOT test conditions to form compounds which may lead to tube or filter screen deposits.

What to Do:
- Perform ASTM D-1319, GC-MS or some other type of analysis to determine whether fuel contains higher molecular weight aromatics and/or naphthenics.

Potential Solution:
- Hydroprocess to reduce fuel aromatic and/or naphthenic-aromatic compounds.

Fuel Contains Dissolved Copper
Copper can catalyze fuel degradation under the JFTOT conditions and cause failure.

What to Do:
- Analyze jet fuel for the presence of copper by ICAP or AA.

Potential Solutions:
- Utilize a metal chelating additive to help minimize the effect of copper catalyzed degradation.
- Filter fuel through clay or molecular sieve to remove copper.

PROBLEM: JET FUEL FAILS PARTICULATE CONTAMINATION TEST
Identifying Possible Causes

Corrosion Inhibitor Has Reacted with Caustic
During the processing of fuel, caustic washing and sweetening are often utilized to remove naphthenic acids, mercaptans and hydrogen sulfide. If not completely removed, caustic can carry over into the fuel and initiate problems. Caustic can effectively react with and neutralize carboxylic acid based corrosion inhibitors added to the jet fuel. As a result, the corrosion inhibitor becomes ineffective at

preventing corrosion and improving lubricity. Also, the salt of the corrosion inhibitor can plug fuel filters to restrict fuel flow.

Test method ASTM D-2276, Particulate Contaminant in Aviation Fuel by Line Sampling, requires filtration of fuel through a 0.8-micron filter. Caustic neutralized corrosion inhibitor salts can block these filters and slow down the filtration of fuel enough to cause failure of this ASTM test.

What to Do:
- Determine whether caustic water is present in the fuel storage or processing system by checking the pH of the storage tank water bottoms.
- If pH is >7, caustic carryover is possible.
- Analyze for the presence of sodium or calcium in the jet fuel. If present, water contamination is likely.

Potential Solutions:
- Maintain schedule of frequent removal of water bottoms from the jet fuel storage system.
- Water wash caustic treated jet fuel to remove traces of caustic.
- After caustic washing or water washing, ensure that adequate settling time has been allowed before the addition of corrosion inhibitors and other additives to finished jet fuel.

PROBLEM: FUEL CORRODES COPPER, BRONZE OR BRASS COMPONENTS
Identifying Possible Causes
Fuel Contains Hydrogen Sulfide or "Active" Sulfur
Copper and copper containing alloys are susceptible to attack by elemental sulfur and hydrogen sulfide as well as organically bound sulfur. "Active" or elemental sulfur and hydrogen sulfide gas can attack copper to form copper sulfide, a dark brown to black compound. This can be seen through ASTM D-130 copper corrosion testing.

What to Do:
- Titration for hydrogen sulfide.
- Evaluate for elemental sulfur.

Potential Solutions:
- Caustic wash to remove hydrogen sulfide.
- Treat with sulfur scavenger/sulfur neutralizer.
- Protect copper with copper corrosion inhibitor.

Fuel Contains Dissolved Sulfuric Acid

Sulfuric acid used in the processing of fuels can carry over into finished products. It can attack copper and copper-containing alloys to form copper sulfate, a blue-green deposit.

What to Do:
- Extract acidic components from fuel with distilled water and check for decrease in pH.

Potential Solution:
- Lightly caustic wash fuel to remove sulfuric acid. The sodium sulfate salt which forms is water soluble and is removed from the fuel.

CHAPTER 9

Components of Fuel and Fuel Additive Storage and Injection Systems

A. METALS

1. Steel

Carbon steel is the most common metal used throughout fuel storage, handling and transportation systems. The term *carbon steel* describes steel containing carbon as the principal alloying element added to control hardness and strength. The carbon content can range from approximately 0.05% to about 1.4%. Above 1.4%, steel becomes brittle and loses cohesiveness and hardness.

Slowly cooled steels are relatively soft. Rapidly quenched steels are strong and hard. Information on the typical uses of steel with varying levels of carbon appear in **TABLE 9-1**.

TABLE 9-1. Uses for Steel with Varying Carbon Content

Percent Carbon	Typical Use
0.05 - 0.12	Nails, wire, chain
0.10 - 0.20	Soft, tough steel. Structural steel, case hardened machine parts, screws
0.20 - 0.30	Better grade of machine and structural steel. Gears, shafts, levers
0.30 - 0.40	Responds to heat treatment. Connecting rods, shafts, axles
0.40 - 0.50	Crankshafts, gears, axles
0.60 - 0.70	Tool steel. Set screws, locomotive wheels, screw drivers
0.70 - 0.80	Tough, hard steel. Anvils, hammers, cable wire, wrenches

0.80 - 0.90	Metal punches, rivet sets, cold chisels
0.90 - 1.00	Provides hardness and tensile strength. Springs, axes, various dies
1.00 - 1.10	Drills, taps, milling tools, knives
1.10 - 1.20	Used where toughness is the most important consideration. Ball bearings, drills, woodworking tools
1.20 - 1.30	Files, knives, tools for milling brass and wood
1.25 - 1.40	Used where sharp cutting edge is important. Razors, saws, wear resistant machine parts

A four or five digit numbering system has been established to classify and identify different types of steel. Originally developed by the Society of Automotive Engineers (SAE) this system is now used by the American Iron and Steel Institute (AISI) to classify steel.

In the numbering system, the first digit identifies the type of steel. The second digit is used to identify the approximate content of the primary alloying element. The remaining two digits indicate the typical carbon content percentage. In the case of certain corrosion resistant and heat resistant steels, a three-digit system is used to identify the average carbon content percent.

In addition, a prefix may be used to identify the process used to manufacture the steel. For example:

A = a basic open-hearth alloy steel
B = an acid-Bessemer carbon steel
C = a basic open-hearth carbon steel
D = an acid open-hearth carbon steel
E = an electric furnace steel

Examples of types of SAE and AISI steel are provided in **TABLE 9-2**.

TABLE 9-2. Types of SAE and AISI Steel

Steel Type	Identification
Plain Carbon Steel	10XX
Resulfurized Carbon Steel	11XX
Resulfurized, Rephosphorized Carbon Steel	12XX
Manganese Steel	13XX
High Manganese Carburizing Steel	15XX
Nickel Steel	2XXX
5% Nickel Steel	25XX
Nickel-Chromium Steel	3XXX
3.5% Nickel, 1.5% Chromium Steel	33XX
Corrosion and Heat Resistant Nickel Steel	30XXX
Carbon-Molybdenum Steel	40XX
Chromium-Nickel-Molybdenum Steel	43XX
Chromium Steel	5XXX
Low Chromium Steel	51XX
Corrosion and Heat Resistant Chromium Steel	51XXX
Chromium-Vanadium Steel	6XXX
Nickel-Chromium-Molybdenum Steel	86XX and 87XX
Manganese-Nickel-Chromium-Molybdenum Steel	94XX

a. Impurities in Steel

Solid impurities such as iron oxides, manganese oxide, silicon and aluminum can become trapped in the steel and form points of weakness in the steel. Weak points can be the site of eventual wear, fracture or corrosion.

Gaseous impurities such as nitrogen, hydrogen, oxygen and carbon monoxide are always present. When present as bubbles, they can decrease the plasticity and increase the embrittlement of steel.

b. Alloying Elements

The addition of low concentrations of elements to steel such as manganese, titanium, boron can greatly enhance the properties of steel. Improved hardness, strength, machinability and resistance to corrosion can all be improved by alloying. The effect of various alloying elements is provided in **TABLE 9-3**.

TABLE 9-3. Effect of Alloying Elements on Steel Properties

Alloying Element	Performance Property
Boron	Increases hardness
Chromium	Improves abrasion and wear resistance
Cobalt	Increases hardness
Copper	Improves atmospheric corrosion resistance
Lead	Reduces cutting friction during machining
Manganese	Promotes forgeability
Molybdenum	Improves hardness, heat treatability and corrosion resistance
Nickel	Increases toughness
Phosphorus	May weaken steel
Sulfur	Forms iron sulfide readily and results in brittle steel
Titanium	Increases strength and heat resistance

2. Stainless Steel

Stainless steel is quite useful as a construction material due to its resistance to corrosion and scaling at high temperatures. The performance of stainless steel is provided by adding chromium, nickel, molybdenum or other alloying elements to steel during processing.

A minimum chromium concentration of approximately 11% is typical for stainless steel. As more chromium is added, corrosion resistance improves. Concentrations of chromium >20% are found in some alloys. Chromium addition leads to the formation of a tight forming oxide film on the surface of the metal. This stable film is "self-healing," which means that the film will reform if scratched or broken. This oxide is quite resistant to attack by acids, bases, organic compounds and inorganic salts.

Nickel is added to improve resistance to environmental and stress cracking corrosion. Molybdenum provides even greater resistance to corrosion and improves mechanical strength. Copper increases resistance to sulfuric acid attack.

Examples of common stainless steel grades and their properties are provided in **TABLE 9-4.**

Components of Fuel and Fuel Additive Storage and Injection Systems

TABLE 9-4. Common Stainless Steel Grades and Their Properties

Stainless Steel Type	Alloying Element	Applications
301	Cr & Ni at low levels	General utility steel for structural uses and household utensils
304 & 304L*	Ni higher levels	Improve welding properties and workability in construction
316 & 316L*	Mo for corrosion resistance	Improve tensile strength of constructed materials

* Most commonly used for piping materials.

3. Copper and Its Alloys

Copper is softer and more ductile than steel and is utilized frequently in the manufacture of pipes and tubing. Copper has good corrosion resistance but will corrode in the presence of nitric acid and other mineral acids. Organic acids do not corrode copper as readily. Dry ammonia does not corrode copper, but the presence of water in ammonia and ammonium hydroxide will corrode copper. Copper resists corrosion in the presence of caustic solutions, but the addition of zinc will increase corrosion rates. Also carbonate, phosphate and silicate salts of sodium will corrode copper.

FIGURE 9-1. Reactions of Ammonia with Copper

$$Cu + 4NH_3 \rightarrow Cu(NH_3)_4^{++} + 2e^-$$

$$2Cu(NH_3)_4^{++} + H_2O + 2e^- \rightarrow Cu_2O + 2NH_4^+ + 6NH_3$$

Copper and its non-ferrous metal alloys, bronze and brass, are used to manufacture tubing, ferrules, valves and a variety of fittings. Although their use is somewhat limited in automotive fuel systems, they are found commonly throughout fuel storage and distribution systems. Copper steam coils and brass hardware may be utilized due to their excellent resistance to corrosion and high level of thermal conductivity. Described below are some of the more common alloys of copper and their applications:

a. Brass

Admiralty Brass
Composition: 70% copper, 29% zinc, 1% tin and 0.03% arsenic. It has good corrosion resistance, especially in sea water. Arsenic inhibits loss of zinc from the alloy. It is used for tubing applications in condensers, preheaters, evaporators and heat exchangers which contact salt water, oil, steam and other liquids below 500°F.

Muntz Metal
Composition: 60% copper and 40% zinc. It is used in the manufacture of screws, valve stems, brazing rods, condenser tubes, condenser heads and heat exchangers baffles and plates. Muntz metal is strong, ductile and corrosion resistant.

Forging Brass
Composition: 60% copper, 38% zinc, 2% lead. Used in hardware and plumbing parts due to its good hot working properties.

Architectural Bronze
A brass composed of 57% copper, 40% zinc, 3% lead. It is used in decorative molding, hinges, locks and industrial forgings.

Naval Bronze or Tobin Bronze
A brass composed of 60% copper, 39.5% zinc, 0.05% tin. It is resistant to salt water spray and is used in condenser plates, welding rods, propeller shafts and marine hardware.

Manganese Bronze
A brass composed of 58.5% copper, 39% zinc, 1.4% iron, 1% tin and 0.1% manganese. It is strong and has excellent wear resistance. Uses include clutch disks, pump rods, valve stems and welding rods.

Red Brass
Composition: 85% copper and 15% zinc. Commonly found in radiator cores, plumbing pipe, heat exchangers and valves.

Cupronickel Alloys
These copper alloys can contain from 10% to 30% nickel. They are excellent heat exchanger tubes because of their resistance to salt water corrosion at high temperatures.

Aluminum Brass
Composition: 78% copper, 20% zinc, 2% aluminum, 0.03% arsenic. It has better corrosion resistance than admiralty brass due to an ability to "self-heal" when abraded. It is used in applications where high velocities are common such as in marine and land power stations as well as sewage and waste water treatment plants.

b. Bronze

This term was originally used to describe copper-tin alloys. It is now used to describe any copper alloy, except copper-zinc alloys, containing up to 25% of the principal alloying element. Some examples include:

Copper-Tin Bronze
Alloys containing principally copper and tin are referred to as true bronzes. Tin increases hardness and wear resistance more than zinc. It also improves salt water corrosion resistance. Common uses of copper-tin bronze are described in TABLE 9-5.

TABLE 9-5. Common Uses of Copper-Tin Bronze Alloys

% Tin	Use
8	Wire, coins, bronze sheeting
8 - 12	Gears, machine parts, bearings, marine fittings
12 - 20	Principal metal for bearings
20 - 25	Bells

Phosphor Bronze
Phosphorus improves casting of molten metal. For this reason it is used in the production of gears and bearings.

Leaded Bronze
Lead addition improves machinability and sliding-wear properties of bearings. It is usually less than 2% of the bronze composition.

Aluminum Bronze
The concentration of aluminum in this alloy is usually between 5% and 11% of the bronze. Other elements such as iron, silicon, nickel and manganese may be added to increase strength. Aluminum bronze has superior corrosion resistance and wear resistance over tin bronze. Uses include non-sparking hand tools, valve seats and guides, bushings, bearings, decorative grills, imitation gold jewelry and paint pigment.

Silicon Bronze
The silicon content in this bronze is about 4%. Silicon bronze has excellent resistance to corrosion by organic acids and sulfite solutions. Common uses include electrical fittings, boilers, pumps, shafts and marine hardware.

4. Aluminum

Aluminum is a soft, ductile and relatively inexpensive metal. The surface of aluminum readily oxidizes in the air and water to form a highly resistant oxide film. This oxide film serves to make aluminum resistant to attack when used in environments containing sulfides, sulfur dioxide, carbon dioxide and other corrosive gases. It is highly resistant to water initiated corrosion, but is susceptible to galvanic corrosion by trace amounts of copper, tin, lead, nickel or carbon steel. The reaction of aluminum in water to form Bayerite is shown in **Figure 9-2**.

FIGURE 9-2. Reaction of Aluminum in Water to form Aluminum Oxide Trihydrate or Bayerite

$$2Al + 6H_2O \rightarrow Al_2O_3 \cdot 3H_2O + 3H_2 \uparrow$$

Aluminum also has a high degree of thermal and electrical conductivity. At low temperatures, the impact strength of aluminum increases, and for this reason, aluminum is commonly used in cryogenic applications.

Alloys are frequently used to improve the strength and hardness of aluminum. These alloys are resistant to corrosion by most organic compounds including hydrocarbon fuels, mercaptans, aldehydes, ethers, ketones and organo sulfur and nitrogen compounds. They are also resistant to most organic acids, phenols, halogenated organics and alcohols. However, aluminum alloys are sometimes corroded by mixtures of organics.

Inorganic acids, bases and salts with pH values <4 and >9 are not compatible with aluminum. Pitting corrosion of aluminum can occur in contact with salt water and copper halide salts.

$$Al + NaOH + H_2O \rightarrow NaAlO_2 + 1.5H_2 \uparrow$$

The effect of pH on corrosion of alloy aluminum is shown in **Figure 9-3**.

Components of Fuel and Fuel Additive Storage and Injection Systems 219

FIGURE 9-3.

Effect of pH on corrosion of 1100-H14 alloy (aluminum) by various chemical solutions. Observe the minimal corrosion in the pH range of 4–9. The low corrosion rates in acetic acid, nitric acid, and ammonium hydroxide demonstrate that the nature of the individual ions in solution is more important than the degree of acidity or alkalinity. (*Courtesy of Alcoa Laboratories; from* Aluminum Properties and Physical Metallurgy, ed. John E. Hatch, *American Society for Metals, Metals Park, Ohio, 1984, Figure 19, page 295.*)

(H. M. Herro and R. D. Port, *The Nalco Guide to Cooling Water System Failure Analysis,* McGraw-Hill, 1993.)

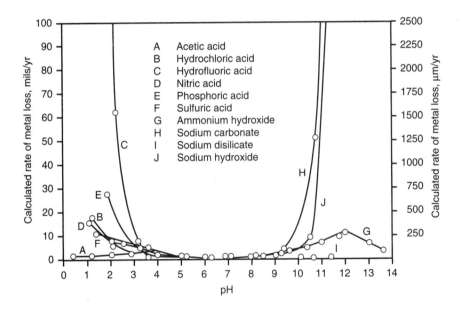

Some of the common aluminum alloying elements and their imparted properties are outlined in **TABLE 9-6.**

TABLE 9.6. Common Aluminum Alloying Elements and Their Properties

Alloying Element	Properties Provided
Copper	Minimizes shrinkage; promotes hardness
Silicon	Improves casting qualities; increases corrosion resistance and impacts toughness
Magnesium	Improves machinability; improves resistance to corrosion by salt water and alkaline solutions
Zinc	Improves mechanical properties

Aluminum cladding is an alloying process used to help prevent surface corrosion of other underlying metal components. This process involves hot rolling metal to produce a protective aluminum barrier. Clad aluminum alloys can be found in some heat exchanger tubing applications.

5. Nickel

Nickel is used throughout industry because of its excellent corrosion resistance. In addition to its use as a steel cladding material to provide corrosion resistance to tanks and production vessel surfaces, nickel is used as an alloying element in steel production. Nickel is resistant to attack by NaOH and other alkali solutions, but is not compatible with ammonium hydroxide. Nickel is resistant to corrosion by sodium chloride solutions, but is corroded severely by iron, copper and mercury chloride salts. Also, nickel has excellent corrosion resistance to most organic acids.

Some of the common nickel alloys are described below:

a. Nickel-Copper (Monel)

Contains approximately 65% nickel, 28% copper and other elements including iron, manganese and cobalt. Monel is very hard, strong and resistant to corrosion. Uses include pumps, valves and chemical industry storage and containment vessels.

b. Nickel-Iron

Contains levels of nickel which vary widely in content. Commercial products include *Invar* and *Permalloy*. These alloys are used in certain high level technology applications such as transoceanic cabling and other applications whereby machining tolerances are exact such as clocks and variable condensers.

c. Nickel-Iron-Molybdenum

These alloys have a high level of resistance to corrosion, especially at elevated temperatures. Commercial products include *Hastalloy A* and *Hastalloy B*. The composition of Hastalloy A and Hastalloy B is given in **TABLE 9-7**.

TABLE 9-7. Composition of Hastelloy A and Hastelloy B

Alloy	% Nickel	% Iron	% Molybdenum	% Others
Hastelloy A	57	20	20	3
Hastelloy B	62	5	30	3

B. PLASTICS AND ELASTOMERS

A variety of natural and synthetic materials are used throughout fuel and lubricant systems. Examples include transfer lines, hoses, fan blades, impellers, small gears, housings and a host of supporting framework. Some plastics can be degraded by fuels, lubricants, additives and various petroleum based compounds. The most resistant material is polytetrafluoroethylene (PTFE). Ryton and Viton are less resistant, but still quite stable in fuel and lubricant systems. Characteristics of PTFE and Ryton are shown below:

PTFE
- Trade names Teflon, Kynar, Halar, Tefzel
- Suitable for use from -200°C to +200°C
- Extreme resistance to fuel combustion by-products and chemical degradation

Ryton
- Resists degradation by ozone
- Resists swelling and leaching by fuel components
- Is flame retardant
- Useful from -40°C to +230°C

The primary considerations for the use of plastic materials in fuel and oil systems include resistance to swelling, softening or embrittlement. Also important are flexibility and stability under a wide range of temperatures. Information on some common elastomers and plastics used in fuel systems is provided in **TABLE 9-8**.

TABLE 9-8. Common Fuel System Elastomers

Trade Name	Chemical Type	Typical Properties	Use
Viton	Fluoro-elastomer	Resistant to degradation by fuel, temperature to 500°F (260°C) and compression	Fuel line housings; metering or needle valves
Lupolen	HDPE	Flexible, and resistant to degradation by fuel and elements of the environment	Fuel tanks
Hypalon	Chlorosulfonated polyethylene	Resistant to fuel and temperatures up to 260°F (126.6°C); flexible at low temperatures	Fuel lines
Zytel	Polyamine	Stable to temperatures up to about 325°F (163°C); resistant to impact, deformation and fatigue	Fuel rails; filter housings

C. CHEMICAL STORAGE AND INJECTION EQUIPMENT

The equipment used to store blend fuels and inject additives into fuel can be designed to fit a specific application. The equipment selection is based upon several factors including the characteristics of the fuel or additive, operating specifications and various regulatory issues.

A typical system includes the following:
- Storage tank and site gauge
- Injection/metering pump
- Transfer lines and hoses
- Flow meter and controller
- Auxiliary equipment such as a heater, insulating blanket or spill containment enclosure

1. Chemical Storage Tanks

Tank configuration can be either vertical or horizontal and can vary in size from 500 gallons to over 1,000,000 gallons. Most tanks are constructed from carbon

Components of Fuel and Fuel Additive Storage and Injection Systems 223

steel although stainless steel, aluminum, fiberglass and various plastics are sometimes used.

Often special requirements or guidelines affect storage tank selection. These can include:

- Vents and vapor recovery systems
- Use of chemical containment area
- Permits
- Location and space
- Exterior painting and coating
- Labeling and identification

Underwriters Laboratories also approves and rates storage tanks. Factors including tank wall thickness, type of metal used, welding procedure, venting, supports, pressure ratings and other related topics are specifically defined in the approval codes. Guidelines which relate directly to the storage of flammable and combustible materials are provided under UL Code 142.

2. Pumps

A variety of positive and non-positive displacement pumps are available for use in fuel and fuel additive systems. Selection is based on application, required flow and cost. The primary differences in positive displacement and non-positive displacement pumps are described in **TABLE 9-9.**

TABLE 9-9. Properties and Performance of Positive Displacement and Non-Positive Displacement Pumps

Positive Displacement	Non-Positive Displacement
Provides a smooth, continuous flow	Delivers fluid as a pulse with each pump stroke or each time the pump chamber outlet opens
Delivery can be reduced or even halted by high pressure; liquid begins to recirculate within the pump	Pressure increase is usually not a problem unless the pressure exceeds the rated capacity of the fittings or gaskets, then leakage can occur
Not effective in systems where pumps are connected in series; pressure differences can affect efficiency	Better in series with other pumps; not as sensitive to pressure differences
Not self-priming; the pump must be started with the housing full of fluid and free of air	Usually self-priming

a. Non-Positive Displacement Pumps

Non-positive displacement pumps operate by centrifugal force. Fluid supplied to the center of a rotating impeller is forced to the outer impeller housing by centrifugal force. The housing is spiral in shape and increases in diameter toward the outlet. The force of the rotating impeller moves the fluid out of the housing outlet at a higher pressure than originally delivered.

A propeller pump is another type of positive displacement pump. It operates similar to a fan blade inside of a pipe or tube. As fluid moves past the blade, the energy of the moving blade delivers fluid through the outlet at a higher pressure than delivered.

b. Positive Displacement Pumps

Vane Pump

This pump contains a slotted rotor driven by a rotating driveshaft. Hardened and polished vanes slide in and out of the slots in the rotor as the rotor turns within an elliptical or circle-shaped cam ring. The spaces between the vanes function as chambers to carry the fluid to be pumped.

As the fluid enters into each vane pump chamber through the pump inlet, the area between the sliding pump vanes fills with fluid. As the rotor continues to move, the chambers carry fluid from the inlet to the pump outlet. The fluid is then squeezed or forced from each individual chamber through the pump outlet at a higher pressure than delivered.

Vane pumps generally have good efficiency and durability. They are effective within a wide pressure range, displacement volume and speed.

A typical vane pump is shown in **FIGURE 9-4.**

FIGURE 9-4. Vane Pump Cross Section

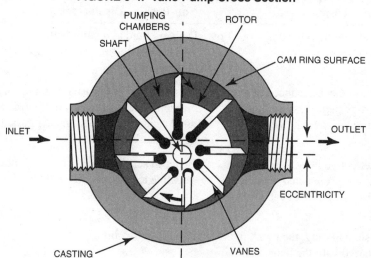

Gear Pump

Gear pumps have a large range of applications. They can be found in automatic transmission systems and in power steering units. The gear pump consists of two interlocking, surface hardened gears which revolve to move fluid through the gear teeth at varying rates of speed. One gear is driven by a power source while the interlocking gear is not. Its rotation is controlled by the powered gear.

As the gear teeth rotate and travel past the pump inlet, a partial vacuum forms. Oil is carried by the small chambers formed between the gear teeth. As the teeth mesh near the pump outlet, oil is delivered at a higher pressure. Also, flow from a gear pump is continuous and not intermittent.

Due to their design, gear pumps are not typically used to meter small quantities of fluid. This is due to the fact that the pump cannot compensate for pressure changes or flow rate changes readily. Gear pumps are better designed for high speed and high flow rates. These pumps are durable and effective at varying pressures. However, their accuracy as a metering pump is limited.

A typical gear pump is shown in **FIGURE 9-5**.

FIGURE 9-5. Gear Pump Cross Section

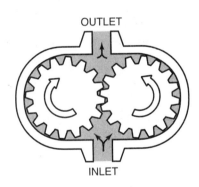

Rotor or Lobe Pump

This type of pump operates in a fashion similar to a gear pump. External gears control the operation of lobe shaped impellers within the pump housing. Displacement by this type of pump is much higher than a gear pump, but the potential for wear is also greater. These pumps are typically used to move large volumes of a fluid at low pressures.

A typical rotor or lobe pump is shown in **FIGURE 9-6**.

FIGURE 9-6. Rotor or Lobe Pump Cross Section

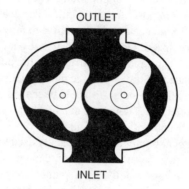

Piston Pump

The piston pump consists of a plunger (piston) which moves within a cylinder-like compartment. The path length of the moving piston can be adjusted and/or the piston speed can be adjusted to control flow rates. These pumps are good at controlling flow at varying pumping rates and operate well at high flow rates and pressures. Pressures as high as 5000 psi are possible.

Piston pumps are reliable and durable but are usually higher in cost than other pumps. Examples of piston pump types include radial piston pumps, axial piston pumps, swash plate in-line piston pumps, wobble plate in-line piston pump and bent axis or angle type pumps. A piston pump and component parts are shown in **FIGURE 9-7.**

Figure 9-7. Piston Pump

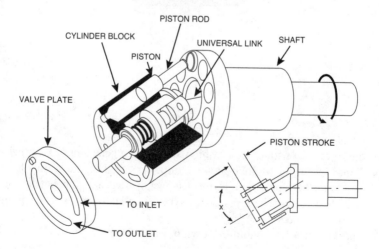

Diaphragm Pump
The diaphragm pump consists of a piston which pushes at varying stroke lengths on an oil reservoir. The hydraulic pressure of this oil then moves a flexible, yet rigid disk-like diaphragm. The diaphragm is typically composed of Teflon. The movement of this diaphragm acts to either create a vacuum or exert positive pressure on a fuel or chemical on the opposite side of the diaphragm. This pulsing effect acts to move fluids through at intermittent, controlled rates.

Diaphragm pumps are used for metering small amounts of additive into a fuel or fuel oil. The cost of these pumps is low compared to other positive displacement pumps. These pumps are excellent metering pumps and are primarily designed for low pressure, low flow applications. Also, they are not recommended for pumping high viscosity fluids.

Air Driven Pump
These pumps usually require a high level of maintenance, but are low in cost. Their performance is affected by moisture, temperature changes and pressure changes.

3. Flow Meters

These units sense the flow of a liquid by translating mechanical energy of moving gears or pistons into an electrical signal. This signal is relayed to an electronic sensing device which converts the signal into a flow rate. Factors which influence the selection of a flow meter include the fluid flow rate, viscosity and compatibility with materials of construction.

Flow meters can be classified as either *positive displacement or as turbine* type meters. Listed below is a brief description of each flow meter type:

a. Positive Displacement Type Flow Meters

Oscillating Piston
- Limited to 150 psi
- Tolerant of dirt and debris
- Good for viscous fluids

Oval Gear
- Extremely accurate
- Tolerant of dirt and debris in fuel
- Good for viscous fluids
- Good for high temperature and pressure applications

Piston Type
- Cannot tolerate dirt and debris in fuel stream

- Excellent for viscous fluids
- Effective at high pressures
- Not recommended for temperatures >200°F

b. Turbine Type Flow Meters

In-Line Turbine
- Not recommended for viscous fluids
- High level of tolerance to dirt and debris
- Good for high temperature and pressure applications
- Very accurate at high and low flow rates

Insertion Turbine
- Identical to "in-line" meters but much larger in size

Insertion Paddlewheel
- Not accurate at low flow rates
- Not recommended for viscous fluids
- Tolerant of dirt and debris
- Low cost

4. Transfer Lines and Hoses

When transferring fuel and fuel additives from storage to the pump and to its final point of use, different types of lines and hoses are utilized. Mild steel, stainless steel and cross-linked polyethylene hose are the typical materials of construction used. These materials withstand the temperature and environmental extremes of outdoor installation and are compatible with fuels and most additives.

Often it is necessary to heat transfer lines to ensure that products do not gel in the line during cold weather. Both electrical heating and steam heating can be used to "trace" lines. *Only explosion proof* electrical heat tape and thermostats can be used on fuel systems. Immersion heaters are not recommended for fuel and fuel additive applications.

If electrical heat tracing is used, it should be attached only to the bottom of the transfer line. *Never wrap a fuel or chemical line with electrical heat tracing. Severe overheating can occur.*

Steam tracing is accomplished by winding copper tubing around the length of pipe to be heated. Steam heat is *very intense* and must be carefully installed to ensure that overheating of piping does not occur. Steam tracing installations are usually permanent, whereby electrical tracing has a shorter life span.

5. Electrical Service

Positive displacement pumps can be built to run on 440-volt three-phase service, 220-volt service or 110-volt single-phase service. The site of electrical service may be restricted in some fuel production and distribution areas. Often, explosion proof enclosures and associated equipment must be used when locating equipment in a hazardous area. If explosion proof equipment is required, the cost of the equipment will increase significantly.

D. VEHICLE FUEL TANKS

The fuel tank must be resistant to corrosion, mechanical impact, temperature fluctuations and internal pressure increases. Caps must not permit fuel to escape when the vehicle is turning, traveling up and down steep grades or when subjected to minor jolts. Also, the tank must be fitted with appropriate relief valves to permit the escape of excessive pressure.

Tank construction materials include steel, aluminum and plastic. Steel tanks are typically lined internally with a corrosion resistant material. Internal coating materials have included Terne (a tin-lead alloy), zinc containing compounds and organic coatings. Aluminum tanks are light in weight and resistant to internal corrosion but are sensitive to external corrosion by salts. Plastic tanks are produced from high density polyethylene and medium density polyethylene. They are highly corrosion resistant and can be molded into various sizes and shapes.

To help ensure resistance to impact and internal pressure increases, plastic tank wall thickness is greater than steel tank thickness. Plastic tanks are typically >4 mm in thickness, whereas steel tanks are somewhat less than 1 mm in thickness. Plastic tanks, however, are considerably lighter in weight.

E. DIESEL FUEL FILTERS

The function of the diesel fuel filter is to prevent contaminants from entering the high pressure fuel pump and causing damage to sensitive pump parts. Fuel filters are designed to trap organic sludge, wax and particulate matter. They also help to coalesce water from fuel into larger droplets which then fall and are collected on the bottom of the filter. Also, some filters are equipped with electronic water sensing devices and bottom drains for removing water.

Although there is no worldwide standard under which fuel filters are manufactured, most contain a hydrophobic paper element with a porosity ranging from approximately 5 to 15 microns with an average porosity of about 8 microns. The porosity of filters used in distributor type fuel pumps is smaller and generally ranges from 4 to 5 microns.

The paper elements can be folded into various forms which appear as pleated, star-shaped or "V"-shaped patterns. These elements are wrapped or coiled around a central porous metal core. Fuels generally flow from the outside shell of the filter, inward through the filter media, and up through the central core of the filter to the fuel pump feed line.

Other filter designs such as the stacked-disk filter exist, but are not common.

F. DIESEL FUEL INJECTION PUMPS

A typical fuel injection system is composed of the following components: fuel tank, injection pump, mechanical or electronic governor, timing device, fuel supply pump, high pressure fuel injection lines and fuel injectors. The high pressure fuel injection pump is core of the fuel injection system. The in-line pump and distributor pump are the most common in fuel systems.

1. Distributor Pump or Rotary Fuel Pump

A typical distributor pump or rotary fuel pump consists of both low pressure and high pressure fuel delivery components. The low pressure delivery components consist of a vane-type fuel supply pump, fuel pressure control valve and a fuel overflow-return valve. The high pressure fuel delivery components consist primarily of a distributor-plunger drive shaft mechanism, a rotating and reciprocating metering plunger and a stationary distributor head through which fuel lines run to individual delivery valves and injectors. The metering plunger has a series of slots and passages along its length for fuel to enter and exit. The distributor head contains ports and channels which enable fuel to pass into the high pressure pump chamber at low pressure and then out of the high pressure pump chamber at much greater pressure.

Fuel to be injected first flows from the pump interior and into the high pressure injection pump chamber through a fuel inlet port in the distributor housing. The reciprocating plunger retracts along its stroke and unblocks the fuel inlet port to enable fuel to flow freely into the high pressure pump chamber. On its return stroke, the fuel inlet port is again blocked by the plunger. The fuel now contained in the high pressure pump chamber then fills the passages in the plunger.

As the metering plunger rotates and lines up with the correct discharge port in the distributor head, fuel is forced from the passages, through a slot in the plunger body and into the proper injection line via the reciprocating action of the plunger. The exiting fuel enters the appropriate line in the distributor head at high pressure. The high injection pressure of the fuel opens the delivery valve and pressurized fuel flows to the fuel injector.

As the reciprocating plunger continues along its path, slots in the plunger open into the low pressure environment of the pump interior. Pressurized fuel remaining in the plunger passageway exits through the slots and into the interior of the pump. The rotating plunger then reverses its stroke path and fuel once again enters the high pressure pump chamber through the fuel inlet port.

This fuel distribution and injection process occurs four times per plunger revolution for a four-cylinder engine and six times per revolution for a six-cylinder engine. Distributor pumps are not typically utilized on eight-cylinder or larger diesel engines.

The amount of fuel metered is adjusted and controlled by the engine governor. All parts of the distributor pump are contained in one housing. Distributor pumps are lubricated by fuel which fills the housing.

High pressure diesel fuel pumps are not tolerant of dirt, debris and organic deposits and can be seriously damaged by fuel contaminants. The barrel and plunger clearance is within a 1-2 micron tolerance. This tolerance is necessary to ensure that fuel injection pressures are maintained with minimal leakage past the plunger shaft. For this reason, thorough and complete filtration of fuel is required before the fuel reaches the injection pump.

2. In-Line Injection Pumps

Standard in-line fuel injection pumps are driven by a camshaft following the speed of the engine. For 2-cycle engines, the camshaft speed is identical to the crankshaft RPM. For 4-cycle engines, the camshaft speed is one-half of the crankshaft RPM. The in-line injection pump parts, such as the camshaft and tappets, are lubricated by the engine oil.

A separate plunger and barrel is utilized for injection of fuel into each cylinder. For example, the in-line pump for a six-cylinder diesel engine contains six individual plunger and barrel assemblies within the pump housing. Plunger movement is controlled by a tappet and spring mechanism driven by the pump camshaft. The plunger movement in the barrel provides the pumping action needed to inject fuel at high pressure into the fuel lines and through the fuel injectors. The machining tolerance between the plunger and barrel assembly is so tight that no additional seal is needed to prevent fuel from leaking past the plunger as it moves within the barrel sleeve.

The process of fuel injection first involves movement of fuel from the fuel tank by the fuel delivery pump. This low pressure pump supplies fuel into the fuel delivery port of each plunger and barrel assembly. As the plunger moves past and clears the fuel delivery port, fuel flows into the barrel filling the chamber above the top of the plunger. Continuous rotation of the camshaft then forces the plunger in the opposite direction, closing off the flow of fuel into the barrel and trapping fuel within the barrel chamber.

The fuel now contained within the barrel is moved by the force of the plunger stroke to the delivery valve. The delivery valve is held in a closed position by a high tension spring. As the plunger stroke continues, the fuel hydraulic pressure increases. Once the hydraulic pressure exceeds the force exerted by the high tension spring on the delivery valve, the valve opens. Fuel then flows at high pressure through the delivery valve and into the fuel line and to the fuel injector. The pressure of the injected fuel then enables the fuel injector to deliver fuel at high pressure into the combustion chamber.

Honed into the side of the plunger is a groove or slot, often termed the "helical groove." This groove is open to the top of the plunger and is filled with fuel. Further down the shaft of the plunger, the groove is aligned with the fuel outlet port, or spill port, of the barrel. As soon as the groove opens to the spill port, the pressure of the injected fuel instantly drops, the fuel delivery valve closes and fuel delivery stops.

CHAPTER 10

Safe Shipping and Hazard Information for Common Fuels, Oils and Solvents

Information is sometimes needed about the human and environmental hazards associated with handling and shipping various petroleum products. This information can often be obtained by utilizing the physical property values and product safety data information compiled for hazardous materials.

This chapter contains valuable safety and hazard information for common fuels, oils and solvents.

A. HAZARDOUS MATERIAL SHIPPING GUIDELINES

Common carriers require the following information on hazardous materials before shipment:

    A. PROPER SHIPPING NAME
    B. HAZARD CLASS OR DIVISION
    C. UN OR I.D. NUMBER
    D. PACKING GROUP
    E. SUBSIDIARY RISK

These requirements are briefly discussed below:

1. PROPER SHIPPING NAME

Proper description of the material being shipped is the first step which must be taken to ensure that fuels and oils are handled, shipped, stored and used properly. A listing of the most commonly accepted names for a variety of petroleum products is provided in **TABLE 10-1**.

Use of the designation "n.o.s." or "not otherwise specified" follows some of the proper shipping names provided in TABLE 10-1. This designation and shipping name can be used *if the material to be shipped contains 25% or less of the hazardous substance.* If the material contains greater than 25% of a single hazardous substance, the "n.o.s." designation may not be appropriate.

2. Hazard Class or Division

Non-Hazardous Materials

Many materials may be classified as non-hazardous materials such as residue, scale, sludge, water and metal parts. These materials may be shipped as *Non-Hazardous Materials*.

Hazardous Materials

Presently, there are nine hazard classes under which hazardous materials are shipped. These are described as follows:

- CLASS 1. Explosives
- CLASS 2. Gas
- CLASS 3. Flammable Liquid
- CLASS 4. Flammable Solids
- CLASS 5. Oxidizing Substances and Organic Peroxides
- CLASS 6. Toxic and Infectious Substances
- CLASS 7. Radioactive Material
- CLASS 8. Corrosives
- CLASS 9. Miscellaneous Dangerous Goods

Most fuels and oils are classified as *Flammable Liquids*. Occasionally *Gas* samples and *Flammable Solids* are shipped. Further details describing each of the nine hazard classes are provided in TABLE 10-1.

3. UN or I.D. Number

The UN or I.D. Number is specific for each material listed under the hazardous materials regulation. TABLE 10-1 contains UN or I.D. Numbers for many common petroleum products.

4. Packing Group

Identification of the correct Packing Group helps to ensure that the proper shipping container is used to transport hazardous materials. Selection of the proper packing group also helps to minimize the risk associated with shipping hazardous materials.

Most hazardous materials are assigned to one of the following packing groups according to the level of danger they present:

- Packing Group I — great danger
- Packing Group II — medium danger
- Packing Group III — minor danger

5. Subsidiary Risk

Some materials may possess more than one hazard risk. As an example, methanol is considered a *Hazard CLASS 3, Flammable Liquid*. It also has a *SUBSIDIARY*

RISK which falls in a *Hazard CLASS 6.1, Toxic Substance*. Subsidiary Risks are listed in **TABLE 1**.

6. Packaging

Regulatory labels describing the hazard characteristics of a fuel or oil sample are required on the outside of boxes and containers used for shipping samples. The labels identify the samples as one or more of the following:

- Flammable
- Oxidizer
- Toxic
- Corrosive

Hazardous Material Labeling Guidelines

a) The hazard labels must be placed on the box in an upright position and must not be covered by any other type of marking or label.
b) If a subsidiary risk exists for a material, the subsidiary risk label must be placed near, but not on, the primary hazard label.
c) If the proper shipping name and UN number are not included on the label, this information must be printed on the box near the label.
d) Also, before air-shipping, an orange colored *CARGO AIRCRAFT ONLY* label must be applied to one side of the shipping container.

7. INTERNATIONAL AND DOMESTIC CARGO SHIPPING CARRIERS

Cargo carriers sometimes have special requirements pertaining to shipment of hazardous materials. To assist customers, most of these carriers provide 24-hour telephone support to help answer questions about cargo shipments.

TABLE 10-1. Hazardous Material Shipping Information
(*Taken from International Air Transport Association Dangerous Goods Regulations, 37th Edition 1996*)

UN or ID#	PROPER SHIPPING NAME/DESCRIPTION	Haz. Class	Sub Risk	Hazard Label	Pack Grp.
1987	Alcohols, n.o.s.	3		Flamm. Liquid	II

UN or ID#	PROPER SHIPPING NAME/DESCRIPTION	Haz. Class	Sub Risk	Hazard Label	Pack Grp.
2734	Amines, liquid, corrosive, flammable	8	3	Corros. & Flamm. Liquid	II
1719	Caustic alkali liquid, n.o.s.	8		Corros.	II
1136	Coal tar distillates, flammable	3		Flamm. Liquid	II
1956	Compressed gas, n.o.s.	2.2		Non-flamm. Gas	Cannot ship by air
1954	Compressed gas, flammable, n.o.s.	2.1		Flamm. Gas	Cannot ship by air
1760	Corrosive liquid, n.o.s.	8		Corros.	II
2920	Corrosive liquid, flammable, n.o.s.	8	3	Corros. & Flamm. Liquid	II
1759	Corrosive solid, n.o.s.	8		Corros.	I
1202	Diesel fuel	3		Flamm. Liquid	III
2801	Dye, liquid, corrosive, n.o.s.	8		Corros.	II
1170	Ethanol	3		Flamm. Liquid	II
3271	Ethers, n.o.s.	3		Flamm. Liquid	II
1993	Flammable liquid, n.o.s.	3		Flamm. Liquid	II
2924	Flammable liquid, corrosive, n.o.s	3	8	Flamm. Liquid & Corros.	II
3286	Flammable liquid, toxic, corrosive, n.o.s.	3	6.1 8	Flamm. Liquid & Toxic & Corros.	Cannot ship by air
1325	Flammable solid, organic, n.o.s.	4.1		Flamm. Solid	II

Safe Shipping and Hazard Information for Common Fuels, Oils and Solvents

UN or ID#	PROPER SHIPPING NAME/DESCRIPTION	Haz. Class	Sub Risk	Hazard Label	Pack Grp.
1863	Fuel, aviation, turbine engine	3		Flamm. Liquid	II
1202	Gas oil	3		Flamm. Liquid	III
1203	Gasoline	3		Flamm. Liquid	II
1202	Heating oil, light	3		Flamm. Liquid	III
1964	Hydrocarbon gas, compressed, n.o.s.	2.1		Flamm. Gas	Cannot ship by air
1965	Hydrocarbon gas, liquefied, n.o.s.	2.1		Flamm. Gas	Cannot ship by air
3295	Hydrocarbon liquid, n.o.s.	3		Flamm. Liquid	I or II
1223	Kerosene	3		Flamm. Liquid	III
3181	Metal salts of organic compounds, flammable, n.o.s.	4.1		Flamm. Solid	II
1230	Methanol	3	6.1	Flamm. Liquid & Toxic	II
1267	Petroleum crude oil	3		Flamm. Liquid	II
1075	Petroleum gases, liquefied	2.1		Flamm. Gas	Cannot ship by air
1268	Petroleum products, n.o.s.	3		Flamm. Liquid	I or II
2735	Polyamines, liquid, corrosive, n.o.s.	8		Corros.	I or II
3244	Solids containing corrosive liquid, n.o.s.	8		Corros.	II
3175	Solids containing flammable liquid, n.o.s.	4.1		Flamm. Solid	II
1999	Tars, liquid	3		Flamm. Liquid	II

B. HAZARD INFORMATION FOR COMMON FUELS AND REFINED PRODUCTS

GASOLINE

CAS Number:	8006-61-9
UN Number:	UN 1203/UN 1257
Classification:	Flammable liquid
Hazard Class:	3
Description:	Clear, aromatic, volatile liquid; a mixture of aliphatic and aromatic hydrocarbons
Flash Point:	-50°F (45.5°C)
AIT:	Approximately 600°F (315°C)
Boiling Points:	IBP variable from 35°F to 102°F (1.7°C to 38.9°C); EP approximately 435°F (223.9°C)
Solubility:	Insoluble in water; freely soluble in alcohol, ether, chloroform and benzene
TSCA:	Reported on inventory
OSHA PEL:	TWA 300 ppm; STEL 500 ppm
Safety Profile:	Mildly toxic by inhalation. Human systemic effects by inhalation include coughing, eye irritation, hallucinations or poor perception. Prolonged skin exposure can cause dermatitis and blistering of the skin. Prolonged inhalation of vapors can result in depression of the central nervous system. Aspiration can cause severe pneumonitis.
	Gasoline is a suspected carcinogen. The vapors are considered to be moderately poisonous. Gasoline can react vigorously with oxidizing materials, heat or flame.
Fire Fighting:	Use dry chemical, foam or CO_2

Examples of hazardous compounds which may be found in gasoline are listed in **TABLE 10-2.**

TABLE 10-2. Examples of Hazardous Compounds Typically Found in Gasoline

Gasoline Component	Approximate %	CAS Number
Benzene	1-5	71-43-2
Cyclohexane	2	110-82-7
Ethylbenzene	1-3	100-41-4
n-Hexane	2-5	110-54-3
Methyl Tertiarybutyl Ether	0-15	1634-04-4
Toluene	7-14	108-88-3
Trimethylbenzene Isomers	5	25551-13-7
2,2,4-Trimethylpentane	3-10	540-84-1
Xylene	8-15	1330-20-7

KEROSENE/JET A

CAS Number: 8008-20-6
UN Number: UN 1223
Classification: Flammable liquid
Hazard Class: 3
Description: Pale-yellow to water-white, hydrocarbon liquid; a mixture of petroleum hydrocarbons having 10-16 carbon atoms per molecule
Flash Point: 100°F to 150°F (37.8°C to 65.5°C)
AIT: 410°F (210°C)
Solubility: Insoluble in water; miscible with other petroleum solvents
TSCA Info: Reported in inventory
Safety Profile: Kerosene is a suspected carcinogen. Ingestion can result in coughing, vomiting, fever, hallucinations and poor perception. Aspiration can result in severe pneumonitis.
Fire Fighting: Use dry chemical, foam or CO_2; vapors are explosive when exposed to heat or flame

DIESEL FUEL OR FUEL OIL

CAS Number:	68476-34-6
UN Number:	NA 1202
Classification:	Flammable liquid
Hazard Class:	3
Description:	A petroleum fraction consisting of a complex mixture of aromatic, paraffinic, olefinic and naphthenic hydrocarbons.
Flash Point:	100°F to 150°F (37.8°C to 65.5°C)
AIT:	495°F (257.2°C) typical
Boiling Point:	300°F to 690°F (149°C to 366°C) typical range
TSCA:	
Safety Profile:	Mildly toxic when ingested.
Fire Fighting:	Use dry chemical or CO_2; combustible when exposed to heat or flame

Examples of hazardous compounds which may be found in diesel fuel are listed in **TABLE 10-3.**

TABLE 10-3. Examples of Hazardous Compounds Typically Found in Diesel Fuel

Diesel Fuel Component	Approximately %	CAS Number
Benzene	<0.1	71-43-2
Cyclohexane	0.1-5	110-82-8
Ethylbenzene	0.1	100-41-4
n-Hexane	<0.1	110-54-3
Mixed Xylenes	<1	1220-20-7
Naphthalene	<1	91-20-3
Toluene	<1	108-88-3
Trimethylbenzene Isomers	<1	25551-13-7

LIGHT VACUUM GAS OIL

CAS Number:	64741-58-8
Classification:	Flammable liquid
Hazard Class:	3
TSCA:	Reported on Inventory
Safety Profile:	Possible carcinogen; pulmonary aspiration can cause severe pneumonitis

HEAVY CATALYTICALLY CRACKED NAPHTHA

CAS Number:	Multiple - Mixture - hydrocarbons C_5 & C_6 with boiling range from 38°C to 220°C (100°F to 430°F)
Hazard Class:	3
Flash Point:	-40°F (-40°C)
Specific Gravity:	0.7 to 0.8
AIT:	Approximately 550°F (400°C)
Safety Profile:	Vapor concentrations >1000 ppm are irritating to the eyes and respiratory tract. Prolonged skin contact will lead to possible irritation and dermatitis
Fire Fighting:	Extinguish with foam, water fog, dry chemical, carbon dioxide and vaporizing liquid fire fighting agents

#6 FUEL OIL/BUNKER C

CAS Number:	68553-00-4
Flash Point:	180°F (82.2°C)
IBP:	400°F (204°C)
Safety Profile:	Vapors can cause respiratory system irritation and dizziness. Direct contact can be severely irritating to eyes and skin

XYLENE

CAS Number:	1330-20-7
UN Number:	UN 1307
Classification:	Flammable Liquid
Hazard Class:	3
TSCA:	Reported on TSCA inventory
Flash Point, PMCC:	83°F (28°C)
OSHA PEL:	TWA 100 ppm; STEL 150 ppm
Vapor Pressure:	6.72 mmHg @ 70°F (21°C)

ETHANOL

CAS Number	64-17-5
UN Number:	UN 1170
Classification:	Flammable Liquid
Hazard Class:	3
TSCA:	Reported in EPA TSCA inventory
Flash Point, PMCC:	55.6°F (13.1°C)
OSHA PEL:	TWA 1000 ppm
Vapor Pressure:	40 mm @ 19°F (-7.2°C)
AIT:	793°F (422.8°C)
Solubility:	Soluble in water, chloroform and ether
Safety Profile:	Confirmed human carcinogen. Moderately toxic by ingestion; mildly toxic by inhalation and skin contact
Fire Fighting:	Use alcohol foam, CO_2 and dry chemical

ETHYLENE GLYCOL MONOMETHYL ETHER

CAS Number:	109-86-4
UN Number:	UN 1188
Classification:	Flammable Liquid
Hazard Class:	6
Flash Point, COC:	115°F (46.1°C)
AIT:	545°F (285°C)
Boiling Point:	124.5°F (51.4°C)
TSCA:	Reported in EPA TSCA Inventory
OSHA PEL:	TWA 25 ppm (skin)

C. HAZARD CLASS DESCRIPTIONS

Nine hazard classes have been identified for describing and grouping hazardous materials. This information may prove useful when advising others on how to properly identify and classify fuels and other hazardous materials.

Class 1 - Explosives

Shipment of explosives by air is not permitted.

Class 2 - Gases

Compressed, Liquefied, Dissolved Under Pressure or Deeply Refrigerated
   This class is comprised of the following:

- Permanent gases - gases which cannot be liquefied at ambient temperature
- Liquefied gases - gases which can become liquid under pressure at ambient temperature
- Dissolved gases - gases dissolved under pressure in a solvent, which may be absorbed in a porous substance
- Deeply refrigerated permanent gases - e.g., liquid air, oxygen, etc.

Class 3 - Flammable Liquids
- These are liquids, mixtures of liquids or liquids containing solids in solution or in suspension. The flash point of these materials is 140°F (60°C) or less by the closed-cup method or 150°F (65.5°C) or less by the open-cup method.

Class 4 - Flammable Solids
These materials are divided into three divisions:
1) Flammable solids - solids, other than those classified as explosives, which are readily combustible, or may contribute to fire through friction during transport.
2) Substances liable to spontaneous combustion - substances which are susceptible to spontaneous heating during transport; substances which heat up in contact with the air and catch fire.
3) Substances which emit flammable gases when in contact with water.

Class 5 - Oxidizing Substances and Organic Peroxides
These materials are divided into two divisions:
1) Substances which may not combust themselves, but by yielding oxygen, may contribute to the combustion of other materials.
2) Organic substances which contain the R-O-O-R functional group and may be considered derivatives of hydrogen peroxide where one or both of the hydrogen atoms have been replaced by organic radicals. Organic peroxides are thermally unstable and may undergo exothermic, self-accelerating decomposition.

Class 6 - Toxic and Infectious Substances
These materials are defined as follows:
Poisonous (toxic) substances - these are likely to either cause death, serious injury or harm to human health if swallowed, inhaled or in contact with the skin.
Infectious substances - these contain viable microorganisms or their toxins which are known, or suspected to cause disease in animals or humans.

Class 7 - Corrosives

These substances, in the event of leakage, can cause severe damage by chemical action when in contact with living tissue. The escape of these substances from its packaging may also cause damage to other cargo or to the transportation unit.

Class 8 - Miscellaneous Dangerous Substances

This class of substances may present a danger during air transport and substances not covered by another class. These include:

> Magnetized material - a material which has a magnetic strength of 0.002 gauss or more at a distance of 7 feet from any point on the surface of the assembled package.
> Other regulated substances - a liquid or solid which is an anesthetic, noxious or other similar properties which could cause extreme annoyance or discomfort to humans.

CHAPTER 11

Fuel Performance Property and Problem Solving Guide

Contained within this chapter is a summary and quick reference guide for use in solving fuel problems. Provided in **SECTION 1** is a comprehensive listing of fuel physical, chemical and performance properties. This information can be utilized to quickly identify the effect certain properties have on fuel performance. **SECTION 2** is a "troubleshooting" or problem solving guide providing information on the possible cause and potential solutions to a variety of common fuel handling and performance problems.

SECTION 1
High Viscosity

Problems associated with difficulty in filtering, mixing and pumping fuel can usually be linked to an increase in fuel viscosity.

Contamination of fuel with high viscosity products or operation at excessively low temperatures can increase the viscosity of fuel and result in fuel pumping problems. Also, if diesel fuel viscosity at 0°F is greater than 45 cSt, fuel pumping problems are likely to occur.

POSSIBLE EFFECTS

- May result in pumping difficulty at low temperatures; wax-related problems may be evident.
- At excessively low temperatures, highly viscous fuel may inadequately fill the pumping chamber of distributor type fuel injection pumps. Poor fuel distribution and engine operation can result.
- Shearing of high viscosity diesel fuel by the fuel injection pump can result in excessive heat buildup and distortion of the pump components.
- Increase in coefficient of friction resulting in a corresponding decrease in energy efficiency is possible.
- Results in poor atomization of diesel fuel when sprayed through an injector; fuel is sprayed as a stream instead of as small droplets; reduced fuel efficiency and power occurs.

- Flame height is lessened in wick-fed kerosene burners; kerosene consumption is decreased.
- Fuel filter throughput is reduced with increasing fuel viscosity.

High Sulfur Content

Fuel sulfur content is being more closely scrutinized with each passing year. It is known that the burning of fuel sulfur can form sulfur dioxide and sulfur trioxide compounds. In combination with water, these sulfur oxides can form acidic compounds.

High sulfur crudes, especially those with a sulfur content >0.5 wt%, complicate refining and are also more expensive to refine. Sulfur and sulfur compounds are corrosive to non-ferrous metals and can decrease the pH of the crude oil. The sulfur content of most crude oil varies from <0.1% to >5%.

In the U.S. and other parts of the world, both low sulfur diesel fuel and high sulfur diesel fuel are being refined. Because fuel sulfur level has been identified as the primary component of fuel emission particulates and "acid rain," sulfur reduction has been mandated and implemented.

POSSIBLE EFFECTS

- Corrosive toward copper, bronze and brass.
- Increase in SO_x during fuel combustion.
- Increase in combustion exhaust odor.
- Potential fuel stability problems especially if sulfur exists in thiophene structure.
- Can improve fuel lubricity if present in the appropriate chemical structure.
- Can degrade almost any elastomer if sulfur exists in the mercaptan form; limited to 50 ppm in aviation jet fuel.
- Combustion products can cause corrosive wear of metal.
- Can lead to an increase in combustion chamber deposits. Under low temperature, low load and speed, stopping and starting driving conditions, moisture condensation can occur; deposits can result.

High Aromatic Content

Monoaromatic, diaromatic and polycyclic aromatic compounds can all be found in fuels. Although concentrations vary in different fuels, the presence of these aromatics can have both desirable and undesirable consequences.

High aromatic content can cause smoke to form during combustion and can lead to carbon deposition in jet engines. A total aromatic content >30% can cause deterioration of aircraft fuel system elastomers and lead to fuel leakage.

POSSIBLE EFFECTS

- Can contribute to low distillate fuel cetane number.
- Increase in heat of combustion or BTU value of heating oil and residual fuel oil.
- Formation of more carbonaceous deposits when burned; increase in smoke and soot.
- Improvement in solvency with increasing aromatic content.
- Increase in gasoline octane number.
- Increase in specific gravity; decrease in °API.
- High naphthalene content of jet fuel is directly related to smoke during combustion.
- Will burn with "reddish" flame in wick-fed kerosene lamps.
- Will burn brilliantly in "blue flame," pressure-fed kerosene lamps.
- Will cause a reduction in the smoke point of kerosene.

High Paraffin Content

Linear, branched and cyclic paraffins all exist in refined fuel. Fuel performance problems can often be directly related to the type and concentration of paraffin present.

Gums, deposits and fuel degradation products will not be dissolved or held in solution by high paraffin content fuels. As a result, gums and degradation products will fall from solution and settle onto fuel system parts such as storage tank bottoms and fuel system lines.

POSSIBLE EFFECTS

- Poor solvent for polar compounds such as fuel additives.
- May contribute to lower specific gravity and BTU values.
- High cetane number can be related to a high diesel paraffin content.
- Increase in the possibility of wax related problems at low temperatures.
- Higher kerosene "smoke point."
- Will burn with yellow flame in kerosene burners.

Low Flash Point

Flash point is considered to be an important specification for all finished fuels and oils. The flammability and combustibility characteristics of a material are directly related to the flash point. Also, fuel transportation codes require flammable compounds to be appropriately labeled for safety reasons.

According to OSHA, compounds with flash point values <100°F are considered flammable. DOT and UN codes rate compounds flammable when the flash point is <141°F.

The flash point, vapor pressure and autoignition temperature values provide important information about the volatility of fuels and solvents. Likewise, the UN number, hazard class and safety profile information of products provide information about the safe shipping and handling of fuels, oils and solvents.

POSSIBLE EFFECTS

- The presence of low molecular weight, low flash point compounds in diesel fuel could lead to a shortening of the fuel ignition delay period. In a diesel engine, this could cause rough running due to early combustion of the low flash point compounds.
- Low autoignition temperature.
- High Reid Vapor Pressure readings are usually linked to low flash point compounds.
- Low molecular weight vapors can cause vapor locking; a critical concern at high altitudes in aviation turbine fuels.

High Specific Gravity/Low °API

The specific gravity of a fuel or oil is a function of the weight per standard volume of the product compared to an equal volume of water. Since aromatic compounds have a greater weight per unit volume than do paraffinic hydrocarbons, the specific gravity of a highly aromatic product would be greater than a paraffinic fuel or oil.

The American Petroleum Institute (API) established an arbitrary rating system to measure the specific gravity of crude oils and petroleum products. This system assigns water the °API value of 10. Products heavier than water have °API gravity values <10. Most petroleum products are lighter in weight than water and have °API gravity values >10. For example, diesel fuel has a typical °API of 30 - 40 and gasoline has a typical °API of 50 - 60. Some asphaltic oils are heavier than water and have °API values <10.

As a general rule, low specific gravity fuel has a lower BTU/gallon rating than high specific gravity fuel. This means that low specific gravity fuel will provide less available energy per gallon of fuel burned.

Because fuel and oil weights vary per standard volume, they should always be handled and sold on a weight basis, not volume basis. If sold on a volume basis, the volume should always be corrected to 60°F by utilizing °API volume reduction tables.

POSSIBLE CAUSES

- High content of aromatic compounds including monoaromatics, diaromatics and polyaromatics.
- May contain high concentration of FCC cycle oil, vacuum gas oil or coker gas oil.

POSSIBLE EFFECTS

- Higher BTU value of fuel.
- High specific gravity fuel may stratify or settle when mixed with certain heavy fuel oils of a lower specific gravity.

High Electrical Conductivity

Water can dissolve a variety of organic and inorganic compounds. The presence of these compounds can increase the ability of water to conduct electrical charge. Because of this fact, it is possible to estimate the total dissolved solids (TDS) in water by measuring the change in its ability to conduct electrical charge.

Electrical conductivity of fuel must be maintained at a minimum and maximum level. Fuel must have adequate conductivity to ensure that static charge does not build up in the fuel.

POSSIBLE CAUSES

- High water content in fuel.
- Dissolved metals or highly polar organic compounds may be present.

POSSIBLE EFFECT

- May interfere with capacitance-type fuel gauges.

Low Viscosity

Low fuel viscosity can be due to the presence of low boiling, low molecular weight compounds in the fuel. Contamination with low boiling compounds such as solvents, gasoline and petroleum naphtha can dramatically reduce the viscosity of distillate fuel and residual fuel oil.

A reduction in both the initial boiling point and end point of a fuel can result in a lowering of viscosity. Accompanying a reduction in these distillation parameters is a consequential increase in the concentration of lower molecular weight and often, more volatile compounds.

If a diesel fuel is low in viscosity due to kerosene dilution or solvent dilution, its lubricity is probably poor. The possibility of wear of the high pressure fuel injection pump parts will increase.

POSSIBLE EFFECTS

- Soft, non-penetrating spray from fuel injectors results in poor mixing with air; combustion quality and power are diminished.
- Readily atomizes into small droplets when sprayed from fuel injector.
- Leakage past diesel fuel injection pump plunger in worn equipment may occur; can result in "hot restart" problems whereby fuel leaks past the injection plunger in hot equipment. Restarting may be impossible until the engine cools down.
- Wear of fuel system components can occur due to poor elastohydrodynamic or mixed lubrication properties.

High Acid Number or Neutralization Number

Fuel acid number or neutralization number is a measure of the acidity or basicity of fuel. The presence of acidic compounds can indicate a fuel stability or oxidative degradation problem. Acidic organic compounds can react with other fuel components to initiate the formation of higher molecular weight, fuel insoluble deposits.

Naphthenic acids and sulfur and nitrogen containing heterocycles and naphthenic-aromatic compounds can all be found at various concentrations in distillate fuel fractions. Degradation through condensation type reactions rather than through radical initiated polymerization is typically more common in distillate fuel. Because of this, degradation products are often quite complex in nature and structurally diverse.

Typically, caustic treatment effectively removes these acidic compounds. However, even after caustic treatment, alkali salts of heavier naphthenic acids may still remain oil soluble. In fuel, these compounds can act as very effective emulsifying agents. Fuel haze and particulate contamination can be due to these acid salts.

Thermal and catalytically cracked gasoline fractions can contain significant concentrations of phenols, and low molecular weight organic acids. All of these compounds can initiate gum formation in gasoline.

POSSIBLE EFFECTS

- Can adversely affect water separation properties of fuel.
- Can stabilize emulsions.
- Enhance gum and deposit formation.

Distillation Profile

- Smoke and odor is minimized in diesel fuel if 50% point is low (<575°F/ 302°C).
- Poor BTU value of diesel if 50% point is <450°F (232.2°C).
- Poor startability if 10% point of diesel is too high.
- Warmup time is increased in diesel engine if a wide boiling range exists between 10% and 50% points.
- Carbon residue and crankcase dilution can increase if 90% point of diesel is high.
- Cetane index is inaccurate if fuels with end points below 500°F are evaluated.
- Wax crystal modifier performance may be poor if T-10 to T-90 temperature range is <200°F (93°C).

High Micro Method, Conradson and Ramsbottom Carbon Values

The amount of carbon present in fuel components can be correlated with a tendency to form deposits in fuel systems. Although the use of various detergent and dispersant additives helps to minimize deposit formation, the carbon residue value is still quite useful.

A high carbon value for gasoline, jet fuel or #2 fuel oil is a good indication that the fuel has been contaminated with residual fuel oil. Heavy streams such as VGO, coker gas oil and #6 fuel oil can contaminate gasoline, jet fuel and diesel fuel. These streams tend to form carbon residue when pyrolyzed and can be identified as fuel contaminants through carbon residue testing.

IMPORTANT FACTORS:

- Measure of all the carbonaceous material remaining after evaporation and pyrolysis of volatile fuel compounds.
- In diesel fuel, the presence of alkyl nitrates causes a higher residue than observed in untreated fuel.
- Can be used to identify whether diesel has been contaminated with residual fuel.
- Carbon deposits cause "hot spots." A temperature gradient exists between the hot spot and the cooler adjacent metal. The result is high metal stress, distortion and cracking.

Diesel Fuel Combustion

Since diesel fuel combustion is an autoignition process, the quality and characteristics of the fuel can have a dramatic impact on the efficiency of engine operation. Ignition delay and cetane number go hand-in-hand to influence the quality of the fuel combustion process. They are the primary engine operability parameters directly related to fuel composition.

It is well known that the boiling point of hydrocarbon molecules will generally increase as the number of carbon atoms increases. It is also known that in order for fuel to burn cleanly and completely, enough oxygen must be present to react with all carbon atoms to form either CO_2 or CO. Therefore, if a fuel contains a greater number of carbon atoms, more oxygen will be required to completely combust this fuel to CO_2 or CO.

Diesel fuel with a high T-50, for example >575°F/302°C, will tend to burn with more smoke, soot and hydrocarbon odor than fuel with a lower T-50. This is basically due to the incomplete combustion and oxidation of a great number of high boiling, high carbon content fuel components in a limited oxygen content environment.

IMPORTANT FACTORS:

- Ignition delay is the period between injection of fuel and ignition; can be controlled by engine design, fuel and air temperature, fuel atomization and fuel composition.
- "Black smoke" is caused by excessive fuel injection volume.
- Knock and vibration can be caused by rapid pressure rise between fuel injection and ignition (during ignition delay).
- Ignition delay is typically shorter than the fuel injection period.
- Straight chain paraffins ignite more readily than branched chain paraffins and aromatics of the same carbon number.
- Fuels with low volatility reduce power output due to poor atomization.
- Use of highly volatile fuels can result in reduced power due to inadequate spray droplet penetration.
- High cetane fuels burn with a shorter ignition delay, low peak pressure, less smoke and odor.
- Smoke formation during combustion can be reduced if the aromatic content of the fuel is reduced.
- Air compression temperatures prior to combustion are approximately 1000°F.
- The greater the carbon number of each hydrocarbon type, the greater the cetane number.

Gasoline Combustion

Volatility is important to consider when discussing the combustibility of gasoline and the burning quality of kerosene. Gasoline volatility is crucial to the combustion process. In order to initiate and ensure smooth combustion of gasoline, volatile compounds such as low molecular weight branched paraffins and aromatics must be present. Factors important to consider concerning gasoline combustion are provided as follows:

IMPORTANT FACTORS:
- No single measurement can be used to characterize antiknock performance. It is a physical and chemical phenomena interrelated with engine design and operating conditions.
- Fuels with low volatility will result in hard starting, slow warmup, poor drivability and unequal distribution to cylinders.
- Low 50% point improves cold weather starting and acceleration; not valid for oxygenate blends, especially ethanol blends.
- High 90% point can contribute to knock resistance.
- High 90% point and end point can lead to poor mixing in combustion chamber, increased hydrocarbon emissions, combustion chamber deposits and crankcase dilution.
- Sulfur and phosphorus can degrade catalytic converter efficiency.

Deposit Analysis

Deposit/Component Analysis	Possible Source of Deposit
- Zn, Ca, P, Mg present in deposit	Indicates the presence of motor oil additives
- Mg:Ca ratio is in the range of 1:3 to 1:5	Indicates the presence of fresh water salts
- Mg:Ca ratio is in the range of 3:1 to 5:1	Indicates the presence of sea water
- Oxyalkylated compounds in deposit	Indicates the presence of water soluble or oil soluble cleaning agents; presence of demulsifying agents
- Polymeric amines in deposit	Indicates the presence of dispersant compounds; gasoline detergents & IVD control additives; fuel stabilizers
- Sodium or calcium salt of carboxylic acid in deposit	Indicates the presence of corrosion inhibitor salt
- High ash content	Indicates the presence of metals and metal salts which can contribute to deposit formation and wear

Deposit/Component Analysis	Possible Source of Deposit
- Vanadium in deposit	Indicates presence of residual oil; vanadium can combine with sodium, potassium or sulfur to form compounds which can cause severe corrosion of engine valves; reaction takes place at temperatures greater than 1100°F
- Sodium at high levels in deposit	Can increase the water solubility of deposits; may indicate the presence of salt water, caustic water

SECTION 2

FUEL PROBLEM SOLVING GUIDE

Problem	Possible Causes	Problem Solutions and Prevention
Difficulty in pumping fuel at low temperatures; wax deposition; wax dropout	- Fuel may be highly paraffinic/waxy - Fuel viscosity may be too high to effectively filter and pump at low temperatures - Product is near or at its pour point - Diesel fuel may have been stored for several days below its cloud point temperature; wax settling has resulted	- Reduce IBP/EP - Blend with low viscosity components - Blend with low molecular weight aromatic components - Treat with wax crystal modifier/dispersant - Heat/insulate fuel system hardware
Reduced power/ combustion quality	- Fuel may be too cold or too viscous to properly atomize into finely dispersed droplets - Distillate fuel may be too dilute and not viscous enough to mix with pressurized air when injected into the fuel combustion chamber - Diesel fuel cetane number may be below 40 - T-50% of diesel <450°F - Gasoline may have lost volatile, low vapor pressure components; high T-50 point gasoline - Fuel injector and/or intake valve deposits	- Check fuel viscosity; viscosity may be too high - Minimize the amount of kerosene blended into No. 2-D fuel; low viscosity fuel - Use diesel fuel with a cetane number of 40 or higher - Check fuel for unusually low BTU rating. Fuel may be diluted with low boiling, light end components - ASTM D-86 profile of diesel to identify the volatility characteristics of the fuel (IBP, T-50, EP) - Replace aged or low vapor pressure gasoline with new fuel having

256 Chapter 11

Problem	Possible Causes	Problem Solutions and Prevention
	may be present in gasoline engine	higher RVP - Use gasoline containing a detergent to help remove deposits from fuel intake system
Poor flame quality of kerosene	- Kerosene may be too viscous due to high EP or due to contamination with high boiling components	- Check fuel viscosity and distillation profile. Fuel consumption would be low if fuel is too viscous or IBP is too high - Use kerosene of proper viscosity and volatility
High gum level/ sediment in gasoline or diesel fuel	- High concentration of olefins or oxygen, sulfur and nitrogen containing compounds in fuel - Fuel may be contaminated with high boiling, high molecular weight compounds - Copper may have catalyzed degradation of fuel - Diesel fuel blend contains high concentration of cycle oil - Inorganic salts of organic compounds may be present in fuel. Waterborne solids may be in fuel	- Minimize the use of high olefin content blend stocks - Minimize exposure of fuel to copper, bronze or brass fuel distribution and storage system hardware - Washing fuel with caustic will remove phenolic and acidic compounds from fuel. Water washing may also remove sediment from fuel - Use of fuel containing effective oxidation inhibiting and metal deactivating compounds can minimize gums and sludge - Look for increased amounts of high molecular weight, high

Problem	Possible Causes	Problem Solutions and Prevention
		boiling components in fuel by ASTM D-86. High boiling compounds may contribute to sediment and gums
Deposits in combustion chamber	- Operating conditions such as low temperature, load and low speed can enhance deposit formation - Combustion of highly aromatic fuel can enhance the formation of carbonaceous deposits - An increase in the concentration of high boiling point components in fuel can enhance combustion chamber deposits; high end point fuel; high D-86 residue fuel - High viscosity lubricating oil bright stock can enhance combustion chamber deposits	- Driving profile may be predominantly city, stop & go, short trip. This may lead to the formation of combustion chamber deposits. - Check oil consumption; excessive amounts of lubricating oil may be entering the combustion chamber - ASTM D-86 T-90 temperature, end point and residue may be high. These high boiling components can contribute to combustion chamber deposits - Use fuel containing a detergent. Some detergents help to minimize combustion chamber deposits
Smoke formation during combustion	- Low temperature, low speed, high load operation can enhance smoke formation in diesel engine - High boiling aromatics in fuel can burn with more smoke (naphtha-	- Use fuel with a lower end point temperature/ low residue - Use of lower aromatic content fuel may help reduce smoke - Too much fuel may be injected into the combustion chamber

Problem	Possible Causes	Problem Solutions and Prevention
	lene compounds can form dense smoke during combustion) - Excessive fuel injection volume can enhance smoke formation	Pump plunger lift may be too great - The addition of cetane improver can sometimes help to reduce exhaust smoke
Ferrous Metal Corrosion	- Contamination of system with process water, sea water, or condensate water - Storage and use of fuel which does not possess corrosion inhibiting properties - Contamination with acidic compounds such as sulfuric acid or hydrofluoric acid alkylation catalyst - Microbial influenced corrosion (MIC)	- Remove as much water as possible from the fuel storage and transportation system - Check pH of tank bottom water for presence of acidic compounds. Remove acidic tank bottom water as soon as possible - Treat system with a ferrous metal corrosion inhibitor - Check for the presence of hydrocarbon-utilizing microbes or sulfate-reducing microbes. Treat fuel/water with a biocide
Emulsion Formation	- Contamination with water containing polar organics or organic salts - Microbial contamination and metabolic products can enhance and stabilize emulsions - Presence of high molecular weight, high viscosity components in fuel may stabilize	- Use a demulsifier to break the emulsion - Heat can be used to break some emulsions - Changing the pH of the water can often break an emulsion - The presence of acid forming and sulfate reducing microorganisms may enhance the formation of emulsions. Treat the fuel with a microbiocide - Formulate fuels with a demulsifier to inhibit emulsion formation

Problem	Possible Causes	Problem Solutions and Prevention
	emulsions - High concentrations of certain fuel additives may enhance the emulsification of water with fuel - Naphthenic acid salts may be present in fuel	- Ensure that surfactant compounds and additives added to fuel prior to or during storage do not contribute to final product emulsion stabilization. This can be done by testing the fuel by ASTM D-1094
High Existent Gum in Gasoline	- Presence of inorganic compounds in fuel can increase existent gums - Inorganic salts of organic compounds can increase existent gum - High residue in ASTM distillation may indicate a possible existent gum problem - Oxidized and polymerized fuel degradation products can increase existent gum	- Removal of existent gums from gasoline cannot be achieved by adding an antioxidant to fuel - The formation of additional gums in gasoline can be minimized by treating fuel with an antioxidant - Washing with water or caustic can be used to reduce and remove existent gum in some fuels - Clay filtration can be used to remove existent gum from fuel
Low Induction Time in Gasoline	- High olefin content; possibly due to addition of FCC gasoline to blend - Acid catalyst from alkylate has induced	- Addition of an antioxidant to fuel will inhibit the fuel oxidation process. Most antioxidants function at treat rates from 5 to 50 ppm - Add 2 ppm of a metal deactivator

Problem	Possible Causes	Problem Solutions and Prevention
	fuel oxidation reaction - Gasoline contains methanol. Methanol reduces the oxidative stability of gasoline	- Water washing or caustic washing will remove acidic and certain polar compounds from fuel. This will help improve fuel oxidative stability - Hydroprocessing will reduce fuel olefin content and improve stability
Fuel is Corrosive Toward Copper	- Hydrogen sulfide/ elemental sulfur may be present in fuel - Fuel contains dissolved ammonia and water - Fuel is contaminated with inorganic acids - Mercaptans attack copper to form gel-like copper mercaptide compounds	- Water wash or caustic wash fuel to remove corrosive acids - Use fuel sweetening additives which chemically react with corroding agent. Use sweetener at treat rate of 1 to 2 times the concentration of the corroding agent - Use a copper corrosion inhibitor. These compounds are typically effective in most fuel applications at treat rates from 5 to 25 ppm
Ferrous Metal Corrosion Inhibitor Does Not Prevent Corrosion	- Corrosion inhibitor treat rate is too low to prevent corrosion - Corrosion inhibitor has been stripped from metal surface by a caustic solution - Scale or deposit on	- Add higher concentrations of corrosion inhibitor - Ensure that caustic carryover into finished fuel does not occur - Use of a caustic resistant corrosion inhibitor may be required

Problem	Possible Causes	Problem Solutions and Prevention
	metal is preventing inhibitor from filming metal - Inorganic acids may be present in storage tank water - Microbial influenced corrosion (MIC)	- Drain water bottoms from storage tank frequently - Run NACE test and look for: 1) emulsification; 2) high or low pH in test water
Cetane Number is Low	- Fuel contains a high percentage of aromatics/ low percentage of paraffins - Fuel could be blended with low cetane components such as FCC cycle oil - Highly aromatic kerosene may have been blended into the fuel	- Adjust the fuel blend to include more paraffinic components - Add cetane improver to boost cetane number - Increasing the fuel IPB and T-50 may increase the cetane number - Minimize the addition of kerosene to a No. 2-D fuel
Distillate Fuel Color Darkens/ Deposits Form	- Fuel contains unstable, reactive components such as quinones, indoles, pyrroles, thiophene or naphthenic acid compounds. These, as well as other components, can react to form color bodies and insoluble, higher molecular weight compounds - FCC cycle oil containing unsaturated compounds are blended into fuel - Vacuum gas oil and light coker gas oil	- Caustic washing can remove some color bodies from fuel - Clay filtration will remove color bodies and high molecular weight compounds from fuel - The addition of fuel stabilizers can help inhibit distillate fuel color degradation - Minimize the use of LCO, VGO, CGO and cracked, unstable components - Minimize the use of fuel streams containing high concentrations of heterocyclic compounds. High levels of heterocycles can yield dark and unstable

Problem	Possible Causes	Problem Solutions and Prevention
	blended into fuel may be contributing to stability problems - Fuel degradation is being catalyzed by copper - Acidic by-products of microbial growth are enhancing fuel degradation	fuels - Evaluate fuel water bottoms for microorganisms. Treat with a biocide if a contamination problem is confirmed - Minimize the use of copper, bronze or brass hardware

References

Chapter 1

Kepner, Charles H. and Benjamin B. Tregoe. 1981. *The New Rational Manager*. Princeton: Princeton Research Press.

Chapter 2

Hengstebeck, R. J. 1959. *Petroleum Processing: Principles and Applications*. New York: McGraw-Hill Book Company.

McPherson, L. J. and M. F. Olive. 1992. Alkylation, isomerization, polymerization, hydrotreatment and sulfur production. In *Modern Petroleum Technology*, ed. G. D. Hobson, pp. 517-575, London: Institute of Petroleum.

McPherson, L. J and M. F. Olive. 1992. Cracking and reforming. In *Modern Petroleum Technology*, ed. G. D. Hobson, pp. 517-575, London: Institute of Petroleum.

Nalco Chemical Company. 1994. *Refining Technology Manual*. Sugar Land, Texas.

Scott, Larry. 1996. CG/MS analysis of polar compounds. Report No. 965391. Houston: Core Laboratories.

Suchanek, Arthur J. 1990. Catalytic routes to low-aromatics diesel look promising. *Oil & Gas Journal,* May 7, 1990:109-119.

Chapter 3

Allenson, Stephan J. 1996. (Internal report) Asphaltenes. Sugar Land, Texas: Nalco/Exxon Energy Chemicals, L.P.

American Society for Testing and Materials. 1989. *ASTM and Other Specifications and Classifications for Petroleum Products and Lubricants*. Philadelphia.

American Society for Testing and Materials. 1996. Marine Fuels: Specifications, testing, purchase and use. 19-21 May, New Orleans.

References

Arco Chemical. 1994. (Factual report) Typical properties of fuel oxygenates in gasoline. Potomac, Maryland: Hart/IRI Fuels Information Services.

Banks, R. E. and P. J. King. 1992. Chemistry and physics of petroleum. In *Modern Petroleum Technology*, ed. G. D. Hobson, pp. 517-575, London: Institute of Petroleum.

Bluth, M. J. 1996. (Internal report) Carbon number characterization of atmospheric tower residual. Sugar Land, Texas: Nalco/Exxon Energy Chemicals, L.P.

Bishop, Geoffrey J. and Cyrus P. Henry, Jr. 1993. Aviation fuels. In *Manual on Significance of Tests for Petroleum Products*. ed. George V. Dyroff, pp. 34-53. Philadelphia: American Society for Testing and Materials.

Boren, D. S. 1994. (Internal report) Diesel fuel cetane number testing. Sugar Land, Texas: Nalco/Exxon Energy Chemicals, L.P.

Boren, D. S. 1994. (Internal report) Diesel fuel lubricity. Sugar Land, Texas: Nalco/Exxon Energy Chemicals, L.P.

Chevron Research and Technology Company. 1990. *Motor Gasolines*. Richmond, Calif.: Chevron.

Dix, R. 1993. Crude oils. In *Manual on Significance of Tests for Petroleum Products*. ed. George V. Dyroff, pp. 1-7. Philadelphia: American Society for Testing and Materials.

Furey, R. L., A. M. Horowitz, and N. J. Schroeder. 1993. Automotive gasoline. In *Manual on Significance of Tests for Petroleum Products*. ed. George V. Dyroff, pp. 24-33. Philadelphia: American Society for Testing and Materials.

Griffith, M. G. and C. W. Sigmund. 1985. Controlling compatibility of residual fuel oils. In *Marine Fuels*, ed. Cletus H. Jones, pp. 227-247, Philadelphia: American Society for Testing and Materials.

Jewitt, C. H., S. R. Westbrook, D. L. Ripley, and R. H. Thornton. 1993. Fuels for land and marine diesel engines and for nonaviation gas turbines. In *Manual on Significance of Tests for Petroleum Products*. ed. George V. Dyroff, pp. 54-68. Philadelphia: American Society for Testing and Materials.

Kueter, Klaus E. Asphaltene precipitation: Cause and effects within Oxy-Peru jungle operation. Bakersfield, Calif.: Occidental Oil and Gas Corporation.

Martin, C. J. and Regina Leonard Gray. 1993. Heating and power generation fuels. In *Manual on Significance of Tests for Petroleum Products.* ed. George V. Dyroff, pp. 69-81. Philadelphia: American Society for Testing and Materials.

Peyla, R. J. 1991. Additives to have key role in new gasoline era. *Oil & Gas Journal,* February 11, 1991:53-57.

Popovich, M. and Carl Hering. 1973. *Fuels and Lubricants.* New York: John Wiley & Sons.

United States Department of Defense. 1995. Military specification: Fuel, naval distillate. MIL-F-16884J.

Waite, R. 1989. Filtration equipment. In *Manual of Aviation Fuel Quality Control Procedures,* ed. Rick Waite, pp. 56-79, Philadelphia: American Society for Testing and Materials.

Winkler, Matthew F. 1985. Shipboard fuel handling and treatment for diesel engines. In *Marine Fuels,* ed. Cletus H. Jones, pp. 154-173, Philadelphia: American Society for Testing and Materials.

Yen, Teh Fu, Wen Hui Wu, and George V. Chilingar. 1984. A study of the structure of petroleum asphaltenes and related substances by infrared spectroscopy. *Energy Sources,* 7(3):203-235.

Chapter 4

Allinger, N. L., et al. 1971. *Organic Chemistry.* New York: Worth Publishers.

American Society for Testing and Materials. 1993. *Annual Book of ASTM Standards,* Volume 05.04. Philadelphia: American Society for Testing and Materials.

American Society for Testing and Materials. 1994. The effect of 24-hour aging in D 97 pour point. Philadelphia: American Society for Testing and Materials.

Boren, D. Scott. 1994. (Internal report) Scuffing BOCLE results. Sugar Land, Texas: Nalco/Exxon Energy Chemicals, L.P.

Coley, T. 1981. Diesel fuel systems for low temperature operability. Paper presented at the Coordinating European Council International Symposium on the Performance of Automotive Fuels and Lubricants, Rome, Italy.

Ethyl Corporation. Diesel fuel additives. Report No. PCD 417872.

Exxon Corporation. 1973. *Tables of Useful Information*.

Gary, J. H. and G. E. Handwerk. 1975. *Petroleum Refining: Technology and Economics*. New York: Marcel Dekker, Inc.

Jones, C. H., Editor. 1985. *Marine Fuels*. Philadelphia: American Society for Testing and Materials.

Nalco Chemical Company. 1996. *Water Chemistry*: Part 1. Naperville, Ill.

Kates, Edgar J. and a team of *POWER* editors. 1964. Oil and gas engines: A *POWER* special report, pp. 2-32.

Stavinoha, L. L. and C. P. Henry, Editors. 1981. *Distillate Fuel Stability and Cleanliness*. Philadelphia: American Society for Testing and Materials.

Sutton, Douglas Leslie. 1986. Investigation into diesel operation with changing fuel property. Report No. 860222. Warrendale, Penn.: Society of Automotive Engineers.

Chapter 5

American Society for Testing and Materials. 1996. *1996 Annual Book of ASTM Standards*. 05.01. West Conshohocken, Penn.

American Society for Testing and Materials. 1996. *Marine Fuels: Specifications, Testing, Purchase and Use*. Proceedings from ASTM Technical Course 19-21 May, New Orleans.

Barry, E. G., L. J. McCabe, D. H. Gerke, and J. M. Perez. 1985. Heavy-duty diesel engine/fuels combustion performance and emissions: A cooperative research program. Report No. 852078. Warrendale, Penn.: Society of Automotive Engineers.

Berridge, S. A. 1992. Finishing processes. In *Modern Petroleum Technology*, ed. G. D. Hobson, pp. 517-575, London: Institute of Petroleum.

Boren, D. S. 1994. (Internal report) Diesel cetane engine testing. Sugar Land, Texas: Nalco/Exxon Energy Chemicals, L.P.

Jewitt, C. H., S. R. Westbrook, D. L. Ripley, and R. H. Thornton. 1993. Fuels for land and marine diesel engines and for nonaviation gas turbines. In *Manual on Significance of Tests for Petroleum Products.* ed. George V. Dyroff, pp. 54-68. Philadelphia: American Society for Testing and Materials.

Lacey, Paul I. and Steven R. Westbrook. 1995. Diesel fuel lubricity. Report No. 950248. Warrendale, Penn.: Society of Automotive Engineers.

Mushrush, George W. 1992. Fuel instability 2: Organo-sulfur hydroperoxide reactions. *Fuel Science and Technology International* 10(10):1563-1600.

Owen, Keith and Trevor Coley. 1995. *Automotive Fuels Reference Book.* 2nd ed. Warrendale, Penn.: Society of Automotive Engineers.

Paramins. 1996. Proceedings from Paramins lubricant seminar 3-5 May, New Brunswick, New Jersey.

Pyburn, C. M., F. P. Cahill and R. K. Lennox. 1978. Sulfur compound interactions on copper corrosion test in propane. *Energy Processing/Canada.* March-April, pp. 60-64.

Russell, Linda A. 1996. (Internal report) Carbon number increase. Sugar Land, Texas: Nalco/Exxon Energy Chemicals, L.P.

Schweitzer, Philip A. 1986. *Corrosion Resistance Tables.* 3d ed. New York: Marcel Dekker.

Stinson, Karl W. 1971. *Diesel Engineering Handbook.* 11th ed. Stamford: Diesel Publications, Inc.

Unzleman, George H. 1993. Cetane index: EPA's control for aromatics. *Octane Week,* 1 August, p. 7.

Yon-Hin, Paul. 1996. (Internal report) Corrosion inhibitors. Sugar Land, Texas: Nalco/Exxon Energy Chemicals, L.P.

Chapter 6

Deen, H. E., A. M. Kaestner, and C. M. Stendahl. 1968. Additives to improve quality and low temperature handling of middle distillate and residual fuels. Published paper presented at World Petroleum Congress, Paris, France.

Dorris, Michelle M. and David Pitcher. 1988. Effective treatment of microbiologically contaminated fuel storage tanks. In *Distillate Fuel: Contamination, Storage and Handling*. ed. Howard L. Chesneau and Michelle M. Dorris, pp. 146-156. Philadelphia: American Society for Testing and Materials.

Herbstman, Sheldon. 1990. *Lubrication*. Vol. 76, No. 2. White Plains, New York: Texaco Inc.

Johnson, J. C. 1975. *Antioxidants: Syntheses and Applications*. Park Ridge, New Jersey: Noyes Data Corporation.

Mikkonen, Seppo, Aapo Niemi, Markku Niemi, and Jorma Niskala. 1989. Effect of engine oil on intake valve deposits. Report No. 89211. Warrendale, Penn.: Society of Automotive Engineers.

Peyla, R. J. 1991. Additives to have key role in new gasoline era. *Oil & Gas Journal*, 11 February, pp. 53-57.

Owen, Keith and Trevor Coley. 1995. *Automotive Fuels Reference Book*. 2nd ed. Warrendale, Penn.: Society of Automotive Engineers.

Ryznar, John W. 1954. Role of chemical additives in the use of lower cost diesel fuel oils. Proceedings from presentation at the Southeastern Railway Diesel Club, August.

Vardi, J. and B.J. Kraus. 1992. Peroxide formation in low sulfur automotive diesel fuels. Report No. 920826. Warrendale, Penn.: Society of Automotive Engineers.

Chapter 7

American Society for Testing and Materials. 1996. *1996 Annual Book of ASTM Standards*. 05.01. West Conshohocken, Penn.

American Society for Testing and Materials. 1996. *1996 Annual Book of ASTM Standards*. 05.02. West Conshohocken, Penn.

American Society for Testing and Materials. 1996. *1996 Annual Book of ASTM Standards*. 05.03. West Conshohocken, Penn.

American Society for Testing and Materials. 1996. *1996 Annual Book of ASTM Standards*. 05.04. West Conshohocken, Penn.

American Society for Testing and Materials. 1995. Standard test method for evaluating lubricity of diesel fuels by the high-frequency reciprocating rig (HFRR). Draft of proposed method. Philadelphia.

American Society for Testing and Materials. 1995. Standard test method for evaluating lubricity of diesel fuels by the scuffing load ball-on-cylinder lubricity evaluator (SLBOCLE). Draft of proposed method. Philadelphia.

Amoco Chemicals Corporation. Pumpability test for distillate fuels. Technical bulletin. Chicago.

Diesel Cold Flow Test Development Group. 1991. Method for determining the simulated filter plugging point of middle distillate fuels. Abingdon, England.

Enjay. Enjay fluidity test. Report No. ELD-59848. Linden, New Jersey.

Institute of Petroleum. 1990. Standard method for analysis and testing of petroleum and related products. London.

NACE International. 1993. Standard test method: Determining the corrosive properties of cargoes in petroleum product pipelines. NACE Standard TM0172-93. Houston.

Nalco Chemical Company. 1984. Procedures for evaluation of pour point depressants. TF-98. Sugar Land, Texas.

Nalco Chemical Company. 1987. Procedures for evaluating the stability of distillate fuel oils. TF-9. Sugar Land, Texas.

STMCS. 1963. Method for determination of Amsterdam maximum pour point.

Texaco. 1988. Instructions for performing the Waring blender test procedure. Beacon, New York.

Universal Oil Products Company. 1974. Doctor test for petroleum distillates. Report No. 41-74.

Williams Brothers Pipeline Company. 1982. Williams Brothers fuel oil stability test.

Chapter 8

Bluth, Matt J. 1997. (Internal report) Low sulfur diesel CFPP reduction. 97/423. Sugar Land, Texas: Nalco/Exxon Energy Chemicals, L.P.

Bourgain, Susan J. 1994. (Internal report) Untreated JP-8 WSIM. 94/417. Sugar Land, Texas: Nalco/Exxon Energy Chemicals, L.P.

Buggs, Ralph N. 1992. (Internal report) Dehazer for gasoline package. 92/496. Sugar Land, Texas: Nalco/Exxon Energy Chemicals, L.P.

Goodell, Mark W. 1991. (Internal report) Untreated LCO stability. 91/397. Sugar Land, Texas: Nalco/Exxon Energy Chemicals, L.P.

Mata, Zoila. 1994. (Internal report) Tank bottom and fuel filter analysis. 94/877. Sugar Land, Texas: Nalco/Exxon Energy Chemicals, L.P.

Nord, Randall F. 1995. (Internal report) Source of ASTM D-2276 failure. 95/693. Sugar Land, Texas: Nalco/Exxon Energy Chemicals, L.P.

Sheets, Sharon O. 1993. (Internal report) Effect of antifoulants on JP-5 WSIM. 93/285. Sugar Land, Texas: Nalco/Exxon Energy Chemicals, L.P.

Sheets, Sharon O. and Randall Nord. 1993. (Internal report) Kerosene quality. 93/45. Sugar Land, Texas: Nalco/Exxon Energy Chemicals, L.P.

Spencer, George O. 1991. (Internal report) Alkylate NACE problem. 91/50. Sugar Land, Texas: Nalco/Exxon Energy Chemicals, L.P.

Wang, Sophia L. 1989. (Internal report) Jet fuel JFTOT/ICAP. 89/434. Sugar Land, Texas: Nalco/Exxon Energy Chemicals, L.P.

Wang, Sophia L. 1991. (Internal report) Diesel calcium content. 91/102. Sugar Land, Texas: Nalco/Exxon Energy Chemicals, L.P.

Wang, Sophia L. 1993. (Internal report) VGO pour point reversion. 93/734. Sugar Land, Texas: Nalco/Exxon Energy Chemicals, L.P.

Wang, Sophia L. 1994. (Internal report) Crude oil pour point reversion. 94/720. Sugar Land, Texas: Nalco/Exxon Energy Chemicals, L.P.

Yon-Hin, Paul. 1995. (Internal report) Naphtha copper corrosion. 95/474. Sugar Land, Texas: Nalco/Exxon Energy Chemicals, L.P.

Chapter 9

Allen, Dell K. 1979. *Metallurgy Theory and Practice*. Chicago: American Technical Society.

Garrett, T. K. 1994. *Automotive Fuels and Fuel Systems*. Vol. 1: *Diesel*. London: Pentech Press.

Garrett, T. K. 1994. *Automotive Fuels and Fuel Systems*. Vol. 2: *Diesel*. London: Pentech Press.

Herro, Harvey M. and Robert D. Port. 1993. *The Nalco Guide to Cooling Water System Failure Analysis*. New York: McGraw-Hill.

Russell, Jim H. 1990. (Internal report) Chemical storage and injection systems. Sugar Land, Texas: Nalco Chemical Company.

Schweitzer, Philip A. 1994. *Corrosion Resistant Piping Systems*. New York: Marcel Dekker.

Sperry Vickers. 1983. *Mobile Hydraulics Manual*. Troy, Michigan: Sperry Corporation.

Tschoke, H. 1994. Distributor injection pumps: VE. In *Diesel Fuel Injection*. ed. Ulrich Adler, Horst Bauer and Anton Beer., pp. 148-161. Robert Bosch GmbH: Stuttgart.

Chapter 10

International Air Transportation Association. 1996. Dangerous goods regulations. Montreal.

Lewis, Richard J. 1993. *Hazardous Chemicals Desk Reference*. 3d ed. New York: Van Nostrand Reinhold.

Mapco Alaska Petroleum. 1996. Material safety data sheet. North Pole, Alaska.

Nalco/Exxon Energy Chemicals, L.P. 1997. *Shipping Samples Safely*. Sugar Land, Texas.

Appendices

Appendix 1. Heat of Combustion of Fuels — Approximate BTU-Gravity Relation

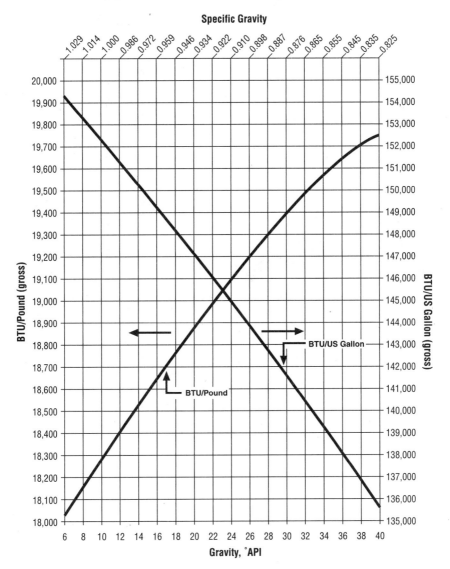

Multiply BTU/pound by 0.556 to obtain kilogram-calories/kilogram
Multiply BTU/gallon by 0.0666 to obtain kilogram-calories/liter

Appendix 2. Factors for Converting Volumes to 60°F

Observed Temperature °F	Group 0 0-14.9° API	Group 1 15.0-34.9° API	Group 2 35.0-50.9° API	Group 3 51.0-63.9° API	Observed Temperature °F	Group 0 0-14.9° API	Group 1 15.0-34.9° API	Group 2 35.0-50.9° API	Group 3 51.0-63.9° API
	Factor for Reducing Volume to 60°F					Factor for Reducing Volume to 60°F			
0	1.0211	1.0241	1.0298	1.0362	40	1.0070	1.0080	1.0099	1.0122
1	1.0208	1.0237	1.0293	1.0356	41	1.0067	1.0076	1.0094	1.0115
2	1.0204	1.0233	1.0288	1.0350	42	1.0063	1.0072	1.0089	1.0109
3	1.0201	1.0229	1.0283	1.0344	43	1.0060	1.0068	1.0084	1.0103
4	1.0197	1.0225	1.0278	1.0338	44	1.0056	1.0064	1.0079	1.0097
5	1.0194	1.0221	1.0273	1.0332	45	1.0053	1.0060	1.0075	1.0091
6	1.0190	1.0217	1.0268	1.0326	46	1.0049	1.0056	1.0070	1.0085
7	1.0186	1.0213	1.0263	1.0320	47	1.0046	1.0052	1.0065	1.0079
8	1.0183	1.0209	1.0258	1.0314	48	1.0042	1.0048	1.0060	1.0073
9	1.0179	1.0205	1.0253	1.0308	49	1.0038	1.0044	1.0055	1.0067
10	1.0176	1.0201	1.0248	1.0302	50	1.0035	1.0040	1.0050	1.0061
11	1.0172	1.0197	1.0243	1.0296	51	1.0031	1.0036	1.0045	1.0054
12	1.0169	1.0193	1.0238	1.0290	52	1.0028	1.0032	1.0040	1.0048
13	1.0165	1.0189	1.0233	1.0284	53	1.0024	1.0028	1.0035	1.0042
14	1.0162	1.0185	1.0228	1.0278	54	1.0021	1.0024	1.0030	1.0036
15	1.0158	1.0181	1.0223	1.0272	55	1.0017	1.0020	1.0025	1.0030
16	1.0155	1.0177	1.0218	1.0266	56	1.0014	1.0016	1.0020	1.0024
17	1.0151	1.0173	1.0214	1.0260	57	1.0010	1.0012	1.0015	1.0018
18	1.0148	1.0168	1.0209	1.0253	58	1.0007	1.0008	1.0010	1.0012
19	1.0144	1.0164	1.0204	1.0247	59	1.0003	1.0004	1.0005	1.0006
20	1.0141	1.0160	1.0199	1.0241	60	1.0000	1.0000	1.0000	1.0000
21	1.0137	1.0156	1.0194	1.0235	61	0.9997	0.9996	0.9995	0.9994
22	1.0133	1.0152	1.0189	1.0229	62	0.9993	0.9992	0.9990	0.9988
23	1.0130	1.0148	1.0184	1.0223	63	0.9990	0.9988	0.9985	0.9982
24	1.0126	1.0144	1.0179	1.0217	64	0.9986	0.9984	0.9980	0.9976
25	1.0123	1.0140	1.0174	1.0211	65	0.9983	0.9980	0.9975	0.9970
26	1.0119	1.0136	1.0169	1.0205	66	0.9979	0.9976	0.9970	0.9964
27	1.0116	1.0132	1.0164	1.0199	67	0.9976	0.9972	0.9965	0.9958
28	1.0112	1.0128	1.0159	1.0193	68	0.9972	0.9968	0.9960	0.9951
29	1.0109	1.0124	1.0154	1.0187	69	0.9969	0.9964	0.9955	0.9945
30	1.0105	1.0120	1.0149	1.0181	70	0.9965	0.9960	0.9950	0.9939
31	1.0102	1.0116	1.0144	1.0175	71	0.9962	0.9956	0.9945	0.9933
32	1.0098	1.0112	1.0139	1.0169	72	0.9958	0.9952	0.9940	0.9927
33	1.0095	1.0108	1.0134	1.0163	73	0.9955	0.9948	0.9935	0.9921
34	1.0091	1.0104	1.0129	1.0157	74	0.9951	0.9944	0.9930	0.9915
35	1.0088	1.0100	1.0124	1.0151	75	0.9948	0.9940	0.9925	0.9909
36	1.0084	1.0096	1.0119	1.0145	76	0.9944	0.9936	0.9920	0.9903
37	1.0081	1.0092	1.0114	1.0139	77	0.9941	0.9932	0.9916	0.9897
38	1.0077	1.0088	1.0109	1.0132	78	0.9937	0.9929	0.9911	0.9891
39	1.0074	1.0084	1.0104	1.0127	79	0.9934	0.9925	0.9906	0.9885

Appendix 2 — Factors for Converting Volumes to 60°F (continued)

Observed Temperature °F	Group 0 0-14.9° API	Group 1 15.0-34.9° API	Group 2 35.0-50.9° API	Group 3 51.0-63.9° API	Observed Temperature °F	Group 0 0-14.9° API	Group 1 15.0-34.9° API	Group 2 35.0-50.9° API	Group 3 51.0-63.9° API
	Factor for Reducing Volume to 60°F					Factor for Reducing Volume to 60°F			
80	0.9930	0.9921	0.9901	0.9879	125	0.9775	0.9744	0.9677	0.9604
81	0.9927	0.9917	0.9896	0.9873	126	0.9771	0.9740	0.9672	0.9598
82	0.9923	0.9913	0.9891	0.9866	127	0.9768	0.9736	0.9667	0.9592
83	0.9920	0.9909	0.9886	0.9860	128	0.9764	0.9732	0.9662	0.9585
84	0.9916	0.9905	0.9881	0.9854	129	0.9761	0.9728	0.9657	0.9579
85	0.9913	0.9901	0.9876	0.9848	130	0.9758	0.9725	0.9652	0.9573
86	0.9909	0.9897	0.9871	0.9842	131	0.9754	0.9721	0.9647	0.9567
87	0.9906	0.9893	0.9866	0.9836	132	0.9751	0.9717	0.9642	0.9561
88	0.9902	0.9889	0.9861	0.9830	133	0.9747	0.9713	0.9637	0.9555
89	0.9899	0.9885	0.9856	0.9824	134	0.9744	0.9709	0.9632	0.9549
90	0.9896	0.9881	0.9851	0.9818	135	0.9740	0.9705	0.9627	0.9542
91	0.9892	0.9877	0.9846	0.9812	136	0.9737	0.9701	0.9622	0.9536
92	0.9889	0.9873	0.9841	0.9806	137	0.9734	0.9697	0.9617	0.9530
93	0.9885	0.9869	0.9836	0.9799	138	0.9730	0.9693	0.9612	0.9524
94	0.9882	0.9865	0.9831	0.9793	139	0.9727	0.9690	0.9607	0.9518
95	0.9878	0.9861	0.9826	0.9787	140	0.9723	0.9686	0.9602	0.9512
96	0.9875	0.9857	0.9821	0.9781	141	0.9720	0.9682	0.9597	0.9506
97	0.9871	0.9854	0.9816	0.9775	142	0.9716	0.9678	0.9592	0.9499
98	0.9868	0.9850	0.9811	0.9769	143	0.9713	0.9674	0.9587	0.9493
99	0.9864	0.9846	0.9806	0.9763	144	0.9710	0.9670	0.9582	0.9487
100	0.9861	0.9842	0.9801	0.9757	145	0.9706	0.9666	0.9577	0.9481
101	0.9857	0.9838	0.9796	0.9751	146	0.9703	0.9662	0.9572	0.9475
102	0.9854	0.9834	0.9791	0.9745	147	0.9699	0.9659	0.9567	0.9469
103	0.9851	0.9830	0.9786	0.9738	148	0.9696	0.9655	0.9562	0.9462
104	0.9847	0.9826	0.9781	0.9732	149	0.9693	0.9651	0.9557	0.9456
105	0.9844	0.9822	0.9776	0.9726	150	0.9689	0.9647	0.9552	0.9450
106	0.9840	0.9818	0.9771	0.9720	151	0.9686	0.9643	0.9547	0.9444
107	0.9837	0.9814	0.9766	0.9714	152	0.9682	0.9639	0.9542	0.9438
108	0.9833	0.9810	0.9761	0.9708	153	0.9679	0.9635	0.9537	0.9432
109	0.9830	0.9806	0.9756	0.9702	154	0.9675	0.9632	0.9532	0.9426
110	0.9826	0.9803	0.9751	0.9696	155	0.9672	0.9628	0.9527	0.9419
111	0.9823	0.9799	0.9746	0.9690	156	0.9669	0.9624	0.9522	0.9413
112	0.9819	0.9795	0.9741	0.9683	157	0.9665	0.9620	0.9517	0.9407
113	0.9816	0.9791	0.9736	0.9677	158	0.9662	0.9616	0.9512	0.9401
114	0.9813	0.9787	0.9731	0.9671	159	0.9658	0.9612	0.9507	0.9395
115	0.9809	0.9783	0.9726	0.9665	160	0.9655	0.9609	0.9502	0.9389
116	0.9806	0.9779	0.9721	0.9659	161	0.9652	0.9605	0.9497	0.9382
117	0.9802	0.9775	0.9717	0.9653	162	0.9648	0.9601	0.9492	0.9376
118	0.9799	0.9771	0.9712	0.9647	163	0.9645	0.9597	0.9487	0.9370
119	0.9795	0.9767	0.9707	0.9641	164	0.9641	0.9593	0.9482	0.9364
120	0.9792	0.9763	0.9702	0.9634	165	0.9638	0.9589	0.9477	0.9358
121	0.9788	0.9760	0.9697	0.9628	166	0.9635	0.9585	0.9472	0.9351
122	0.9785	0.9756	0.9692	0.9622	167	0.9631	0.9582	0.9467	0.9345
123	0.9782	0.9752	0.9687	0.9616	168	0.9628	0.9578	0.9462	0.9339
124	0.9778	0.9748	0.9682	0.9610	169	0.9624	0.9574	0.9457	0.9333

Appendix 2 — Factors for Converting Volumes to 60°F (continued)

Observed Temperature °F	Group 0 0-14.9° API	Group 1 15.0-34.9° API	Group 2 35.0-50.9° API	Group 3 51.0-63.9 API
	Factor for Reducing Volume to 60°F			
170	0.9621	0.9570	0.9452	0.9327
171	0.9618	0.9566	0.9447	0.9321
172	0.9614	0.9562	0.9442	0.9314
173	0.9611	0.9559	0.9437	0.9308
174	0.9607	0.9555	0.9432	0.9302
175	0.9604	0.9551	0.9428	0.9296
176	0.9601	0.9547	0.9423	0.9290
177	0.9597	0.9543	0.9418	0.9283
178	0.9594	0.9539	0.9413	0.9277
179	0.9590	0.9536	0.9408	0.9271
180	0.9587	0.9532	0.9403	0.9265
181	0.9584	0.9528	0.9398	0.9259
182	0.9580	0.9524	0.9393	0.9252
183	0.9577	0.9520	0.9388	0.9246
184	0.9574	0.9517	0.9383	0.9240
185	0.9570	0.9513	0.9378	0.9234
186	0.9567	0.9509	0.9373	0.9228
187	0.9563	0.9505	0.9368	0.9221
188	0.9560	0.9501	0.9363	0.9215
189	0.9557	0.9498	0.9358	0.9209
190	0.9553	0.9494	0.9353	0.9203
191	0.9550	0.9490	0.9348	0.9197
192	0.9547	0.9486	0.9343	0.9190
193	0.9543	0.9482	0.9338	0.9184
194	0.9540	0.9478	0.9333	0.9178
195	0.9536	0.9475	0.9328	0.9172
196	0.9533	0.9471	0.9323	0.9166
197	0.9530	0.9467	0.9318	0.9159
198	0.9526	0.9463	0.9313	0.9153
199	0.9523	0.9460	0.9308	0.9147
200	0.9520	0.9456	0.9303	0.9141
201	0.9516	0.9452	0.9298	
202	0.9513	0.9448	0.9293	
203	0.9509	0.9444	0.9288	
204	0.9506	0.9441	0.9283	
205	0.9503	0.9437	0.9278	
206	0.9499	0.9433	0.9273	
207	0.9496	0.9429	0.9268	
208	0.9493	0.9425	0.9263	
209	0.9489	0.9422	0.9258	

Observed Temperature °F	Group 0 0-14.9° API	Group 1 15.0-34.9° API	Group 2 35.0-50.9° API
	Factor for Reducing Volume to 60°F		
210	0.9486	0.9418	0.9253
211	0.9483	0.9414	0.9248
212	0.9479	0.9410	0.9243
213	0.9476	0.9407	0.9238
214	0.9472	0.9403	0.9233
215	0.9469	0.9399	0.9228
216	0.9466	0.9395	0.9223
217	0.9462	0.9391	0.9218
218	0.9459	0.9388	0.9213
219	0.9456	0.9384	0.9208
220	0.9452	0.9380	0.9203
221	0.9449	0.9376	0.9198
222	0.9446	0.9373	0.9193
223	0.9442	0.9369	0.9188
224	0.9439	0.9365	0.9183
225	0.9436	0.9361	0.9178
226	0.9432	0.9358	0.9173
227	0.9429	0.9354	0.9168
228	0.9426	0.9350	0.9163
229	0.9422	0.9346	0.9158
230	0.9419	0.9343	0.9153
231	0.9416	0.9339	0.9148
232	0.9412	0.9335	0.9143
233	0.9409	0.9331	0.9138
234	0.9405	0.9328	0.9133
235	0.9402	0.9324	0.9128
236	0.9399	0.9320	0.9123
237	0.9395	0.9316	0.9118
238	0.9392	0.9313	0.9113
239	0.9389	0.9309	0.9108
240	0.9385	0.9305	0.9103
241	0.9382	0.9301	0.9098
242	0.9379	0.9298	0.9093
243	0.9375	0.9294	0.9088
244	0.9372	0.9290	0.9083
245	0.9369	0.9286	0.9078
246	0.9365	0.9283	0.9073
247	0.9362	0.9279	0.9068
248	0.9359	0.9275	0.9063
249	0.9356	0.9272	0.9058
250	0.9352	0.9268	0.9053

Appendix 2 — Factors for Converting Volumes to 60°F (continued)

Observed Temperature °F	Group 0 0-14.9° API	Group 1 15.0-34.9° API	Observed Temperature °F	Group 0 0-14.9° API	Group 1 15.0-34.9° API	Observed Temperature °F	Group 0 0-14.9° API	Group 1 15.0-34.9° API
	Factor for Reducing Volume to 60°F			Factor for Reducing Volume to 60°F			Factor for Reducing Volume to 60°F	
251	0.9349	0.9264	295	0.9204	0.9102	340	0.9057	0.8938
252	0.9346	0.9260	296	0.9200	0.9098	341	0.9053	0.8934
253	0.9342	0.9257	297	0.9197	0.9094	342	0.9050	0.8931
254	0.9339	0.9253	298	0.9194	0.9091	343	0.9047	0.8927
			299	0.9190	0.9087	344	0.9044	0.8924
255	0.9336	0.9249						
256	0.9332	0.9245	300	0.9187	0.9083	345	0.9040	0.8920
257	0.9329	0.9242				346	0.9037	0.8916
258	0.9326	0.9238	301	0.9184	0.9080	347	0.9034	0.8913
259	0.9322	0.9234	302	0.9181	0.9076	348	0.9031	0.8909
			303	0.9177	0.9072	349	0.9028	0.8906
260	0.9319	0.9231	304	0.9174	0.9069			
261	0.9316	0.9227				350	0.9024	0.8902
262	0.9312	0.9223	305	0.9171	0.9065	351	0.9021	0.8899
263	0.9309	0.9219	306	0.9167	0.9061	352	0.9018	0.8895
264	0.9306	0.9216	307	0.9164	0.9058	353	0.9015	0.8891
			308	0.9161	0.9054	354	0.9011	0.8888
265	0.9302	0.9212	309	0.9158	0.9050			
266	0.9299	0.9208				355	0.9008	0.8884
267	0.9296	0.9205	310	0.9154	0.9047	356	0.9005	0.8881
268	0.9293	0.9201	311	0.9151	0.9043	357	0.9002	0.8877
269	0.9289	0.9197	312	0.9148	0.9039	358	0.8998	0.8873
			313	0.9145	0.9036	359	0.8995	0.8870
270	0.9286	0.9194	314	0.9141	0.9032			
271	0.9283	0.9190				360	0.8992	0.8866
272	0.9279	0.9186	315	0.9138	0.9029	361	0.8989	0.8863
273	0.9276	0.9182	316	0.9135	0.9025	362	0.8986	0.8859
274	0.9273	0.9179	317	0.9132	0.9021	363	0.8982	0.8856
			318	0.9128	0.9018	364	0.8979	0.8852
275	0.9269	0.9175	319	0.9125	0.9014			
276	0.9266	0.9171				365	0.8976	0.8848
277	0.9263	0.9168	320	0.9122	0.9010	366	0.8973	0.8845
278	0.9259	0.9164	321	0.9118	0.9007	367	0.8969	0.8841
279	0.9256	0.9160	322	0.9115	0.9003	368	0.8966	0.8838
			323	0.9112	0.9000	369	0.8963	0.8834
280	0.9253	0.9157	324	0.9109	0.8996			
281	0.9250	0.9153				370	0.8960	0.8831
282	0.9246	0.9149	325	0.9105	0.8992	371	0.8957	0.8827
283	0.9243	0.9146	326	0.9102	0.8989	372	0.8953	0.8823
284	0.9240	0.9142	327	0.9099	0.8985	373	0.8950	0.8820
			328	0.9096	0.8981	374	0.8947	0.8816
285	0.9236	0.9138	329	0.9092	0.8978			
286	0.9233	0.9135				375	0.8944	0.8813
287	0.9230	0.9131	330	0.9089	0.8974	376	0.8941	0.8809
288	0.9227	0.9127	331	0.9086	0.8971	377	0.8937	0.8806
289	0.9223	0.9124	332	0.9083	0.8967	378	0.8934	0.8802
			333	0.9079	0.8963	379	0.8931	0.8799
290	0.9220	0.9120	334	0.9076	0.8960			
291	0.9217	0.9116				380	0.8928	0.8795
292	0.9213	0.9113	335	0.9073	0.8956	381	0.8924	0.8792
293	0.9210	0.9109	336	0.9070	0.8952	382	0.8921	0.8788
294	0.9207	0.9105	337	0.9066	0.8949	383	0.8918	0.8784
			338	0.9063	0.8945	384	0.8915	0.8781
			339	0.9060	0.8942			

Appendix 2 — Factors for Converting Volumes to 60°F (continued)

Observed Temperature °F	Group 0 0-14.9° API	Group 1 15.0-34.9° API	Observed Temperature °F	Group 0 0-14.9° API	Group 1 15.0-34.9° API	Observed Temperature °F	Group 0 0-14.9° API	Group 1 15.0-34.9° API
	Factor for Reducing Volume to 60°F			Factor for Reducing Volume to 60°F			Factor for Reducing Volume to 60°F	
385	0.8912	0.8777	425	0.8784	0.8637	465	0.8658	0.8498
386	0.8908	0.8774	426	0.8781	0.8633	466	0.8655	0.8495
387	0.8905	0.8770	427	0.8778	0.8630	467	0.8652	0.8492
388	0.8902	0.8767	428	0.8775	0.8626	468	0.8649	0.8488
389	0.8899	0.8763	429	0.8772	0.8623	469	0.8646	0.8485
390	0.8896	0.8760	430	0.8768	0.8619	470	0.8643	0.8481
391	0.8892	0.8756	431	0.8765	0.8616	471	0.8640	0.8478
392	0.8889	0.8753	432	0.8762	0.8612	472	0.8636	0.8474
393	0.8886	0.8749	433	0.8759	0.8609	473	0.8633	0.8471
394	0.8883	0.8746	434	0.8756	0.8605	474	0.8630	0.8468
395	0.8880	0.8742	435	0.8753	0.8602	475	0.8627	0.8464
396	0.8876	0.8738	436	0.8749	0.8599	476	0.8624	0.8461
397	0.8873	0.8735	437	0.8746	0.8595	477	0.8621	0.8457
398	0.8870	0.8731	438	0.8743	0.8592	478	0.8618	0.8454
399	0.8867	0.8728	439	0.8740	0.8588	479	0.8615	0.8451
400	0.8864	0.8724	440	0.8737	0.8585	480	0.8611	0.8447
401	0.8861	0.8721	441	0.8734	0.8581	481	0.8608	0.8444
402	0.8857	0.8717	442	0.8731	0.8578	482	0.8605	0.8440
403	0.8854	0.8714	443	0.8727	0.8574	483	0.8602	0.8437
404	0.8851	0.8710	444	0.8724	0.8571	484	0.8599	0.8433
405	0.8848	0.8707	445	0.8721	0.8567	485	0.8596	0.8430
406	0.8845	0.8703	446	0.8718	0.8564	486	0.8593	0.8427
407	0.8841	0.8700	447	0.8715	0.8560	487	0.8590	0.8423
408	0.8838	0.8696	448	0.8712	0.8557	488	0.8587	0.8420
409	0.8835	0.8693	449	0.8709	0.8554	489	0.8583	0.8416
410	0.8832	0.8689	450	0.8705	0.8550	490	0.8580	0.8413
411	0.8829	0.8686	451	0.8702	0.8547	491	0.8577	0.8410
412	0.8826	0.8682	452	0.8699	0.8543	492	0.8574	0.8406
413	0.8822	0.8679	453	0.8696	0.8540	493	0.8571	0.8403
414	0.8819	0.8675	454	0.8693	0.8536	494	0.8568	0.8399
415	0.8816	0.8672	455	0.8690	0.8533	495	0.8565	0.8396
416	0.8813	0.8668	456	0.8687	0.8529	496	0.8562	0.8393
417	0.8810	0.8665	457	0.8683	0.8526	497	0.8559	0.8389
418	0.8806	0.8661	458	0.8680	0.8522	498	0.8556	0.8386
419	0.8803	0.8658	459	0.8677	0.8519	499	0.8552	0.8383
420	0.8800	0.8654	460	0.8674	0.8516	500	0.8549	0.8379
421	0.8797	0.8651	461	0.8671	0.8512			
422	0.8794	0.8647	462	0.8668	0.8509			
423	0.8791	0.8644	463	0.8665	0.8505			
424	0.8787	0.8640	464	0.8661	0.8502			

Appendix 3. Reduction of Observed API Gravity to API Gravity at 60°F

Observed Temperature °F	API Gravity at Observed Temperature																				
	5	6	7	8	9	10	11	12	13	14	15	16	17	18	19	20	21	22	23	24
	Corresponding API Gravity at 60°F																				
0	7.9	8.9	10.0	11.0	12.1	13.1	14.2	15.2	16.3	17.3	18.3	19.4	20.4	21.5	22.5	23.6	24.6	25.7	26.7	27.8
1	7.8	8.9	9.9	11.0	12.0	13.1	14.1	15.2	16.2	17.2	18.3	19.3	20.4	21.4	22.5	23.5	24.6	25.6	26.7	27.7
2	7.8	8.8	9.9	10.9	12.0	13.0	14.1	15.1	16.1	17.2	18.2	19.3	20.3	21.4	22.4	23.5	24.5	25.6	26.6	27.7
3	7.7	8.8	9.8	10.9	11.9	13.0	14.0	15.0	16.1	17.1	18.2	19.2	20.3	21.3	22.4	23.4	24.4	25.5	26.5	27.6
4	7.7	8.7	9.8	10.8	11.9	12.9	13.9	15.0	16.0	17.1	18.1	19.2	20.2	21.2	22.3	23.3	24.4	25.4	26.5	27.5
5	7.6	8.7	9.7	10.8	11.8	12.8	13.9	14.9	16.0	17.0	18.1	19.1	20.1	21.2	22.2	23.3	24.3	25.4	26.4	27.5
6	7.6	8.6	9.7	10.7	11.8	12.8	13.8	14.9	15.9	17.0	18.0	19.0	20.1	21.1	22.2	23.2	24.3	25.3	26.4	27.4
7	7.5	8.6	9.6	10.7	11.7	12.7	13.8	14.8	15.9	16.9	17.9	19.0	20.0	21.1	22.1	23.2	24.2	25.2	26.3	27.3
8	7.5	8.5	9.6	10.6	11.6	12.7	13.7	14.8	15.8	16.8	17.9	18.9	20.0	21.0	22.0	23.1	24.1	25.2	26.2	27.3
9	7.4	8.5	9.5	10.6	11.6	12.6	13.7	14.7	15.7	16.8	17.8	18.9	19.9	20.9	22.0	23.0	24.1	25.1	26.2	27.2
10	7.4	8.4	9.5	10.5	11.5	12.6	13.6	14.7	15.7	16.7	17.8	18.8	19.8	20.9	21.9	23.0	24.0	25.0	26.1	27.1
11	7.3	8.4	9.4	10.5	11.5	12.5	13.6	14.6	15.6	16.7	17.7	18.8	19.8	20.8	21.9	22.9	23.9	25.0	26.0	27.1
12	7.3	8.3	9.4	10.4	11.4	12.5	13.5	14.5	15.6	16.6	17.7	18.7	19.7	20.8	21.8	22.8	23.9	24.9	26.0	27.0
13	7.2	8.3	9.3	10.3	11.4	12.4	13.5	14.5	15.5	16.6	17.6	18.6	19.7	20.7	21.7	22.8	23.8	24.9	25.9	26.9
14	7.2	8.2	9.3	10.3	11.3	12.4	13.4	14.4	15.5	16.5	17.5	18.6	19.6	20.7	21.7	22.7	23.8	24.8	25.8	26.9
15	7.1	8.2	9.2	10.2	11.3	12.3	13.3	14.4	15.4	16.5	17.5	18.5	19.6	20.6	21.6	22.7	23.7	24.7	25.8	26.8
16	7.1	8.1	9.2	10.2	11.2	12.3	13.3	14.3	15.4	16.4	17.4	18.5	19.5	20.5	21.6	22.6	23.6	24.7	25.7	26.8
17	7.0	8.1	9.1	10.1	11.2	12.2	13.2	14.3	15.3	16.3	17.4	18.4	19.4	20.5	21.5	22.5	23.6	24.6	25.6	26.7
18	7.0	8.0	9.1	10.1	11.1	12.2	13.2	14.2	15.3	16.3	17.3	18.3	19.4	20.4	21.4	22.5	23.5	24.5	25.6	26.6
19	6.9	8.0	9.0	10.0	11.1	12.1	13.1	14.2	15.2	16.2	17.3	18.3	19.3	20.4	21.4	22.4	23.5	24.5	25.5	26.6
20	6.9	7.9	9.0	10.0	11.0	12.0	13.1	14.1	15.1	16.2	17.2	18.2	19.3	20.3	21.3	22.4	23.4	24.4	25.5	26.5
21	6.8	7.9	8.9	9.9	11.0	12.0	13.0	14.1	15.1	16.1	17.1	18.2	19.2	20.2	21.3	22.3	23.3	24.4	25.4	26.4
22	6.8	7.8	8.8	9.9	10.9	11.9	13.0	14.0	15.0	16.1	17.1	18.1	19.1	20.2	21.2	22.2	23.3	24.2	25.3	26.4
23	6.7	7.8	8.8	9.8	10.9	11.9	12.9	13.9	15.0	16.0	17.0	18.1	19.1	20.1	21.1	22.2	23.2	24.2	25.3	26.3
24	6.7	7.7	8.7	9.8	10.8	11.8	12.9	13.9	14.9	15.9	17.0	18.0	19.0	20.1	21.1	22.1	23.1	24.2	25.2	26.2
25	6.6	7.7	8.7	9.7	10.8	11.8	12.8	13.8	14.9	15.9	16.9	17.9	19.0	20.0	21.0	22.1	23.1	24.1	25.1	26.2
26	6.6	7.6	8.6	9.7	10.7	11.7	12.8	13.8	14.8	15.8	16.9	17.9	18.9	19.9	21.0	22.0	23.0	24.1	25.1	26.1
27	6.5	7.6	8.6	9.6	10.7	11.7	12.7	13.7	14.8	15.8	16.8	17.8	18.9	19.9	20.9	21.9	23.0	24.0	25.0	26.0
28	6.5	7.5	8.5	9.6	10.6	11.6	12.7	13.7	14.7	15.7	16.8	17.8	18.8	19.8	20.9	21.9	22.9	23.9	25.0	26.0
29	6.4	7.5	8.5	9.5	10.6	11.6	12.6	13.6	14.6	15.7	16.7	17.7	18.7	19.8	20.8	21.8	22.8	23.9	24.9	25.9
30	6.4	7.4	8.4	9.5	10.5	11.5	12.5	13.6	14.6	15.6	16.6	17.7	18.7	19.7	20.7	21.8	22.8	23.8	24.8	25.9
31	6.3	7.4	8.4	9.4	10.5	11.5	12.5	13.5	14.5	15.6	16.6	17.6	18.6	19.7	20.7	21.7	22.7	23.7	24.8	25.8
32	6.3	7.3	8.4	9.4	10.4	11.4	12.4	13.5	14.5	15.5	16.5	17.6	18.6	19.6	20.6	21.6	22.7	23.7	24.7	25.7
33	6.3	7.3	8.3	9.3	10.4	11.4	12.4	13.4	14.4	15.5	16.5	17.5	18.5	19.5	20.6	21.6	22.6	23.6	24.6	25.7
34	6.2	7.2	8.3	9.3	10.3	11.3	12.3	13.4	14.4	15.4	16.4	17.4	18.5	19.5	20.5	21.5	22.5	23.6	24.6	25.6
35	6.2	7.2	8.2	9.2	10.2	11.3	12.3	13.3	14.3	15.3	16.4	17.4	18.4	19.4	20.4	21.5	22.5	23.5	24.5	25.5
36	6.1	7.1	8.2	9.2	10.2	11.2	12.2	13.3	14.3	15.3	16.3	17.3	18.3	19.4	20.4	21.4	22.4	23.4	24.5	25.5
37	6.1	7.1	8.1	9.1	10.1	11.2	12.2	13.2	14.2	15.2	16.3	17.3	18.3	19.3	20.3	21.3	22.4	23.4	24.4	25.4
38	6.0	7.0	8.1	9.1	10.1	11.1	12.1	13.1	14.2	15.2	16.2	17.2	18.2	19.2	20.3	21.3	22.3	23.3	24.3	25.4
39	6.0	7.0	8.0	9.0	10.0	11.1	12.1	13.1	14.1	15.1	16.1	17.2	18.2	19.2	20.2	21.2	22.2	23.3	24.3	25.3

Appendix 3 — Reduction of Observed API Gravity to API Gravity at 60°F (continued)

Observed Temperature °F	API Gravity at Observed Temperature																				
	5	6	7	8	9	10	11	12	13	14	15	16	17	18	19	20	21	22	23	24
	Corresponding API Gravity at 60°F																				
40	5.9	6.9	8.0	9.0	10.0	11.0	12.0	13.0	14.1	15.1	16.1	17.1	18.1	19.1	20.1	21.2	22.2	23.2	24.2	25.2
41	5.9	6.9	7.9	8.9	9.9	11.0	12.0	13.0	14.0	15.0	16.0	17.0	18.1	19.1	20.1	21.1	22.1	23.1	24.2	25.2
42	5.8	6.8	7.9	8.9	9.9	10.9	11.9	12.9	13.9	15.0	16.0	17.0	18.0	19.0	20.0	21.0	22.1	23.1	24.1	25.1
43	5.8	6.8	7.8	8.8	9.8	10.9	11.9	12.9	13.9	14.9	15.9	16.9	17.9	19.0	20.0	21.0	22.0	23.0	24.0	25.0
44	5.7	6.7	7.8	8.8	9.8	10.8	11.8	12.8	13.8	14.9	15.9	16.9	17.9	18.9	19.9	20.9	21.9	23.0	24.0	25.0
45	5.7	6.7	7.7	8.7	9.7	10.8	11.8	12.8	13.8	14.8	15.8	16.8	17.8	18.8	19.9	20.9	21.9	22.9	23.9	24.9
46	5.6	6.7	7.7	8.7	9.7	10.7	11.7	12.7	13.7	14.7	15.8	16.8	17.8	18.8	19.8	20.8	21.8	22.8	23.8	24.9
47	5.6	6.6	7.6	8.6	9.6	10.7	11.7	12.7	13.7	14.7	15.7	16.7	17.7	18.7	19.7	20.8	21.8	22.8	23.8	24.8
48	5.5	6.6	7.6	8.6	9.6	10.6	11.6	12.6	13.6	14.6	15.6	16.7	17.7	18.7	19.7	20.7	21.7	22.7	23.7	24.7
49	5.5	6.5	7.5	8.5	9.5	10.6	11.6	12.6	13.6	14.6	15.6	16.6	17.6	18.6	19.6	20.6	21.6	22.7	23.7	24.7
50	5.5	6.5	7.5	8.5	9.5	10.5	11.5	12.5	13.5	14.5	15.5	16.5	17.6	18.6	19.6	20.6	21.6	22.6	23.6	24.6
51	5.4	6.4	7.4	8.4	9.4	10.5	11.5	12.5	13.5	14.5	15.5	16.5	17.5	18.5	19.5	20.5	21.5	22.5	23.5	24.6
52	5.4	6.4	7.4	8.4	9.4	10.4	11.4	12.4	13.4	14.4	15.4	16.4	17.4	18.5	19.5	20.5	21.5	22.5	23.5	24.5
53	5.3	6.3	7.3	8.3	9.3	10.4	11.4	12.4	13.4	14.4	15.4	16.4	17.4	18.4	19.4	20.4	21.4	22.4	23.4	24.4
54	5.3	6.3	7.3	8.3	9.3	10.3	11.3	12.3	13.3	14.3	15.3	16.3	17.3	18.3	19.3	20.3	21.4	22.4	23.4	24.4
55	5.2	6.2	7.2	8.2	9.2	10.2	11.3	12.3	13.3	14.3	15.3	16.3	17.3	18.3	19.3	20.3	21.3	22.3	23.3	24.3
56	5.2	6.2	7.2	8.2	9.2	10.2	11.2	12.2	13.2	14.2	15.2	16.2	17.2	18.2	19.2	20.2	21.2	22.2	23.2	24.2
57	5.1	6.1	7.1	8.1	9.1	10.1	11.2	12.2	13.2	14.2	15.2	16.2	17.2	18.2	19.2	20.2	21.2	22.2	23.2	24.2
58	5.1	6.1	7.1	8.1	9.1	10.1	11.1	12.1	13.1	14.1	15.1	16.1	17.1	18.1	19.1	20.1	21.1	22.1	23.1	24.1
59	5.0	6.0	7.0	8.0	9.0	10.0	11.1	12.1	13.1	14.1	15.1	16.1	17.1	18.1	19.1	20.1	21.1	22.1	23.1	24.1
60	5.0	6.0	7.0	8.0	9.0	10.0	11.0	12.0	13.0	14.0	15.0	16.0	17.0	18.0	19.0	20.0	21.0	22.0	23.0	24.0
61	5.0	6.0	7.0	8.0	9.0	10.0	10.9	11.9	12.9	13.9	14.9	16.0	16.9	17.9	18.9	20.0	21.0	21.9	22.9	23.9
62	4.9	5.9	6.9	7.9	8.9	9.9	10.8	11.9	12.9	13.9	14.9	15.9	16.9	17.9	18.9	19.9	20.9	21.9	22.9	23.9
63	4.9	5.9	6.9	7.9	8.9	9.9	10.8	11.8	12.8	13.8	14.8	15.8	16.8	17.8	18.8	19.8	20.8	21.8	22.8	23.8
64	4.8	5.8	6.8	7.8	8.8	9.8	10.8	11.8	12.8	13.8	14.8	15.8	16.8	17.8	18.8	19.8	20.8	21.8	22.8	23.8
65	4.8	5.8	6.8	7.8	8.8	9.8	10.7	11.7	12.7	13.7	14.7	15.7	16.7	17.7	18.7	19.7	20.7	21.7	22.7	23.7
66	4.7	5.7	6.7	7.7	8.7	9.7	10.7	11.7	12.7	13.7	14.7	15.7	16.7	17.7	18.7	19.7	20.7	21.6	22.6	23.6
67	4.7	5.7	6.7	7.7	8.7	9.7	10.6	11.6	12.6	13.6	14.6	15.6	16.6	17.6	18.6	19.6	20.6	21.6	22.6	23.6
68	4.6	5.6	6.6	7.6	8.6	9.6	10.6	11.6	12.6	13.6	14.6	15.6	16.6	17.6	18.5	19.5	20.5	21.5	22.5	23.5
69	4.6	5.6	6.6	7.6	8.6	9.6	10.6	11.5	12.5	13.5	14.5	15.5	16.5	17.5	18.5	19.5	20.5	21.5	22.5	23.5
70	4.5	5.5	6.5	7.5	8.5	9.5	10.5	11.5	12.5	13.5	14.5	15.5	16.5	17.4	18.4	19.4	20.4	21.4	22.4	23.4
71	4.5	5.5	6.5	7.5	8.5	9.5	10.4	11.4	12.4	13.4	14.4	15.4	16.4	17.4	18.4	19.4	20.4	21.4	22.3	23.3
72	4.5	5.4	6.4	7.4	8.4	9.4	10.4	11.4	12.4	13.4	14.4	15.4	16.3	17.3	18.3	19.3	20.3	21.3	22.4	23.3
73	4.4	5.4	6.4	7.4	8.4	9.4	10.3	11.3	12.3	13.3	14.3	15.3	16.3	17.3	18.3	19.3	20.2	21.2	22.2	23.2
74	4.4	5.4	6.3	7.3	8.3	9.3	10.3	11.3	12.3	13.3	14.3	15.2	16.2	17.2	18.2	19.2	20.2	21.2	22.2	23.2
75	4.3	5.3	6.3	7.3	8.3	9.3	10.2	11.2	12.2	13.2	14.2	15.2	16.2	17.2	18.2	19.1	20.1	21.1	22.1	23.1
76	4.3	5.3	6.3	7.3	8.2	9.2	10.2	11.2	12.2	13.2	14.2	15.1	16.1	17.1	18.1	19.1	20.1	21.1	22.0	23.0
77	4.2	5.2	6.2	7.2	8.2	9.2	10.2	11.1	12.1	13.1	14.1	15.1	16.1	17.1	18.0	19.0	20.0	21.0	22.0	23.0
78	4.2	5.2	6.2	7.1	8.1	9.1	10.1	11.1	12.1	13.1	14.0	15.0	16.0	17.0	18.0	19.0	20.0	20.9	21.9	22.9
79	4.1	5.1	6.1	7.1	8.1	9.1	10.1	11.0	12.0	13.0	14.0	15.0	16.0	17.0	17.9	18.9	19.9	20.9	21.9	22.9
80	4.1	5.1	6.1	7.1	8.0	9.0	10.0	11.0	12.0	13.0	13.9	14.9	15.9	16.9	17.9	18.9	19.8	20.8	21.8	22.8
81	4.1	5.0	6.0	7.0	8.0	9.0	10.0	10.9	11.9	12.9	13.9	14.9	15.9	16.8	17.8	18.8	19.8	20.8	21.8	22.7
82	4.0	5.0	6.0	7.0	7.9	8.9	9.9	10.9	11.9	12.9	13.8	14.8	15.8	16.8	17.8	18.8	19.7	20.7	21.7	22.7
83	4.0	5.0	5.9	6.9	7.9	8.9	9.9	10.8	11.8	12.8	13.8	14.8	15.8	16.7	17.7	18.7	19.7	20.7	21.6	22.6
84	3.9	4.9	5.9	6.9	7.8	8.8	9.8	10.8	11.8	12.8	13.7	14.7	15.7	16.7	17.7	18.6	19.6	20.6	21.6	22.6
85	3.9	4.9	5.8	6.8	7.8	8.8	9.8	10.7	11.7	12.7	13.7	14.7	15.6	16.6	17.6	18.6	19.6	20.5	21.5	22.5
86	3.8	4.8	5.8	6.8	7.8	8.7	9.7	10.7	11.7	12.7	13.6	14.6	15.6	16.6	17.5	18.5	19.5	20.5	21.5	22.4
87	3.8	4.8	5.7	6.7	7.7	8.7	9.7	10.6	11.6	12.6	13.6	14.6	15.5	16.5	17.5	18.5	19.5	20.4	21.4	22.4
88	3.7	4.7	5.7	6.7	7.7	8.6	9.6	10.6	11.6	12.5	13.5	14.5	15.5	16.5	17.4	18.4	19.4	20.4	21.3	22.3
89	3.7	4.7	5.7	6.6	7.6	8.6	9.6	10.5	11.5	12.5	13.5	14.5	15.4	16.4	17.4	18.4	19.3	20.3	21.3	22.3

Appendix 3 — Reduction of Observed API Gravity to API Gravity at 60°F (continued)

Observed Temperature °F	API Gravity at Observed Temperature																			
	5	6	7	8	9	10	11	12	13	14	15	16	17	18	19	20	21	22	23	24
	Corresponding API Gravity at 60°F																				
90	3.7	4.6	5.6	6.6	7.6	8.5	9.5	10.5	11.5	12.4	13.4	14.4	15.4	16.4	17.3	18.3	19.3	20.3	21.2	22.2
91	3.6	4.6	5.6	6.5	7.5	8.5	9.5	10.4	11.4	12.4	13.4	14.3	15.3	16.3	17.3	18.3	19.2	20.2	21.2	22.2
92	3.6	4.5	5.5	6.5	7.5	8.4	9.4	10.4	11.4	12.3	13.3	14.3	15.3	16.2	17.2	18.2	19.2	20.1	21.1	22.1
93	3.5	4.5	5.5	6.5	7.4	8.4	9.4	10.3	11.3	12.3	13.3	14.2	15.2	16.2	17.2	18.1	19.1	20.1	21.1	22.0
94	3.5	4.5	5.4	6.4	7.4	8.4	9.3	10.3	11.3	12.2	13.2	14.2	15.2	16.1	17.1	18.1	19.1	20.0	21.0	22.0
95	3.4	4.4	5.4	6.4	7.3	8.3	9.3	10.2	11.2	12.2	13.2	14.1	15.1	16.1	17.1	18.0	19.0	20.0	20.9	21.9
96	3.4	4.4	5.3	6.3	7.3	8.3	9.2	10.2	11.2	12.1	13.1	14.1	15.1	16.0	17.0	18.0	18.9	19.9	20.9	21.9
97	3.4	4.3	5.3	6.3	7.2	8.2	9.2	10.2	11.1	12.1	13.1	14.0	15.0	16.0	17.0	17.9	18.9	19.9	20.8	21.8
98	3.3	4.3	5.3	6.2	7.2	8.2	9.1	10.1	11.1	12.0	13.0	14.0	15.0	15.9	16.9	17.9	18.8	19.8	20.8	21.7
99	3.3	4.2	5.2	6.2	7.1	8.1	9.1	10.1	11.0	12.0	13.0	13.9	14.9	15.9	16.8	17.8	18.8	19.7	20.7	21.7
100	3.2	4.2	5.2	6.1	7.1	8.1	9.0	10.0	11.0	11.9	12.9	13.9	14.9	15.8	16.8	17.8	18.7	19.7	20.7	21.6
101	3.2	4.1	5.1	6.1	7.1	8.0	9.0	10.0	10.9	11.9	12.9	13.8	14.8	15.8	16.7	17.7	18.7	19.6	20.6	21.6
102	3.1	4.1	5.1	6.0	7.0	8.0	8.9	9.9	10.9	11.8	12.8	13.8	14.7	15.7	16.7	17.6	18.6	19.6	20.5	21.5
103	3.1	4.1	5.0	6.0	7.0	7.9	8.9	9.9	10.8	11.8	12.8	13.7	14.7	15.7	16.6	17.6	18.6	19.5	20.5	21.5
104	3.0	4.0	5.0	5.9	6.9	7.9	8.8	9.8	10.8	11.7	12.7	13.7	14.6	15.6	16.6	17.5	18.5	19.5	20.4	21.4
105	3.0	4.0	4.9	5.9	6.9	7.8	8.8	9.8	10.7	11.7	12.7	13.6	14.6	15.6	16.5	17.5	18.4	19.4	20.4	21.3
106	3.0	3.9	4.9	5.9	6.8	7.8	8.8	9.7	10.7	11.6	12.6	13.6	14.5	15.5	16.5	17.4	18.4	19.4	20.3	21.3
107	2.9	3.9	4.8	5.8	6.8	7.7	8.7	9.7	10.6	11.6	12.6	13.5	14.5	15.5	16.4	17.4	18.3	19.3	20.3	21.2
108	2.9	3.8	4.8	5.8	6.7	7.7	8.7	9.6	10.6	11.5	12.5	13.5	14.4	15.4	16.4	17.3	18.3	19.2	20.2	21.2
109	2.8	3.8	4.8	5.7	6.7	7.6	8.6	9.6	10.5	11.5	12.5	13.4	14.4	15.3	16.3	17.3	18.2	19.2	20.2	21.1
110	2.8	3.7	4.7	5.7	6.6	7.6	8.6	9.5	10.5	11.5	12.4	13.4	14.3	15.3	16.3	17.2	18.2	19.1	20.1	21.1
111	2.7	3.7	4.7	5.6	6.6	7.6	8.5	9.5	10.4	11.4	12.4	13.3	14.3	15.2	16.2	17.2	18.1	19.1	20.0	21.0
112	2.7	3.7	4.6	5.6	6.5	7.5	8.5	9.4	10.4	11.4	12.3	13.3	14.2	15.2	16.1	17.1	18.1	19.0	20.0	20.9
113	2.7	3.6	4.6	5.5	6.5	7.5	8.4	9.4	10.3	11.3	12.3	13.2	14.2	15.1	16.1	17.1	18.0	19.0	19.9	20.9
114	2.6	3.6	4.5	5.5	6.5	7.4	8.4	9.3	10.3	11.3	12.2	13.2	14.1	15.1	16.0	17.0	18.0	18.9	19.9	20.8
115	2.6	3.5	4.5	5.5	6.4	7.4	8.3	9.3	10.2	11.2	12.2	13.1	14.1	15.0	16.0	16.9	17.9	18.9	19.8	20.8
116	2.5	3.5	4.4	5.4	6.4	7.3	8.3	9.2	10.2	11.2	12.1	13.1	14.0	15.0	15.9	16.9	17.8	18.8	19.8	20.7
117	2.5	3.4	4.4	5.4	6.3	7.3	8.2	9.2	10.2	11.1	12.1	13.0	14.0	14.9	15.9	16.8	17.8	18.7	19.7	20.7
118	2.4	3.4	4.4	5.3	6.3	7.2	8.2	9.1	10.1	11.1	12.0	13.0	13.9	14.9	15.8	16.8	17.7	18.7	19.6	20.6
119	2.4	3.4	4.3	5.3	6.2	7.2	8.1	9.1	10.1	11.0	12.0	12.9	13.9	14.8	15.8	16.7	17.7	18.6	19.6	20.5
120	2.4	3.3	4.3	5.2	6.2	7.1	8.1	9.1	10.0	11.0	11.9	12.9	13.8	14.8	15.7	16.7	17.6	18.6	19.5	20.5
121	2.3	3.3	4.2	5.2	6.1	7.1	8.1	9.0	10.0	10.9	11.9	12.8	13.8	14.7	15.7	16.6	17.6	18.5	19.5	20.4
122	2.3	3.2	4.2	5.1	6.1	7.0	8.0	9.0	9.9	10.9	11.8	12.8	13.7	14.7	15.6	16.6	17.5	18.5	19.4	20.4
123	2.2	3.2	4.1	5.1	6.0	7.0	8.0	8.9	9.9	10.8	11.8	12.7	13.7	14.6	15.6	16.5	17.5	18.4	19.4	20.3
124	2.2	3.1	4.1	5.0	6.0	7.0	7.9	8.9	9.8	10.8	11.7	12.7	13.6	14.6	15.5	16.5	17.4	18.4	19.3	20.3
125	2.1	3.1	4.0	5.0	6.0	6.9	7.9	8.8	9.8	10.7	11.7	12.6	13.6	14.5	15.5	16.4	17.4	18.3	19.3	20.2
126	2.1	3.0	4.0	5.0	5.9	6.9	7.8	8.8	9.7	10.7	11.6	12.6	13.5	14.5	15.4	16.4	17.3	18.3	19.2	20.2
127	2.1	3.0	4.0	4.9	5.9	6.8	7.8	8.7	9.7	10.6	11.6	12.5	13.5	14.4	15.4	16.3	17.3	18.2	19.2	20.1
128	2.0	3.0	3.9	4.9	5.8	6.8	7.7	8.7	9.6	10.6	11.5	12.5	13.4	14.4	15.3	16.3	17.2	18.2	19.1	20.0
129	2.0	2.9	3.9	4.8	5.8	6.7	7.7	8.6	9.6	10.5	11.5	12.4	13.4	14.3	15.3	16.2	17.2	18.1	19.0	20.0
130	1.9	2.9	3.8	4.8	5.7	6.7	7.6	8.6	9.5	10.5	11.4	12.4	13.3	14.3	15.2	16.2	17.1	18.0	19.0	19.9
131	1.9	2.8	3.8	4.7	5.7	6.6	7.6	8.5	9.5	10.4	11.4	12.3	13.3	14.2	15.2	16.1	17.0	18.0	18.9	19.9
132	1.8	2.8	3.7	4.7	5.6	6.6	7.5	8.5	9.4	10.4	11.3	12.3	13.2	14.2	15.1	16.1	17.0	17.9	18.9	19.8
133	1.8	2.7	3.7	4.6	5.6	6.5	7.5	8.4	9.4	10.3	11.3	12.2	13.2	14.1	15.1	16.0	16.9	17.9	18.8	19.8
134	1.8	2.7	3.7	4.6	5.6	6.5	7.4	8.4	9.3	10.3	11.2	12.2	13.1	14.1	15.0	15.9	16.9	17.8	18.8	19.7
135	1.7	2.7	3.6	4.6	5.5	6.5	7.4	8.4	9.3	10.2	11.2	12.1	13.1	14.0	15.0	15.9	16.8	17.8	18.7	19.7
136	1.7	2.6	3.6	4.5	5.5	6.4	7.4	8.3	9.3	10.2	11.1	12.1	13.0	14.0	14.9	15.8	16.8	17.7	18.7	19.6
137	1.6	2.6	3.5	4.5	5.4	6.4	7.3	8.3	9.2	10.2	11.1	12.0	13.0	13.9	14.9	15.8	16.7	17.7	18.6	19.5
138	1.6	2.5	3.5	4.4	5.4	6.3	7.3	8.2	9.2	10.1	11.0	12.0	12.9	13.9	14.8	15.7	16.7	17.6	18.6	19.5
139	1.5	2.5	3.4	4.4	5.3	6.3	7.2	8.2	9.1	10.1	11.0	11.9	12.9	13.8	14.8	15.7	16.6	17.6	18.5	19.4

Appendix 3 — Reduction of Observed API Gravity to API Gravity at 60°F (continued)

Observed Temperature °F	API Gravity at Observed Temperature																			
	25	26	27	28	29	30	31	32	33	34	35	36	37	38	39	40	41	42	43	44
	Corresponding API Gravity at 60°F																				
0	28.9	30.0	31.0	32.0	33.1	34.2	35.3	36.3	37.4	38.5	39.6	40.7	41.8	42.9	44.0	45.1	46.2	47.3	48.4	49.5
1	28.8	29.8	30.9	32.0	33.0	34.1	35.2	36.3	37.3	38.4	39.5	40.6	41.7	42.8	43.9	45.0	46.1	47.2	48.3	49.4
2	28.7	29.8	30.8	31.9	33.0	34.0	35.1	36.2	37.3	38.3	39.4	40.5	41.5	42.7	43.8	44.9	46.0	47.1	48.2	49.3
3	28.7	29.7	30.8	31.8	32.9	34.0	35.0	36.1	37.2	38.3	39.3	40.4	41.5	42.6	43.7	44.8	45.9	47.0	48.1	49.2
4	28.6	29.6	30.7	31.8	32.8	33.9	35.0	36.0	37.1	38.2	39.3	40.3	41.4	42.5	43.6	44.7	45.8	46.9	48.0	49.1
5	28.5	29.6	30.6	31.7	32.8	33.8	34.9	36.0	37.0	38.1	39.2	40.3	41.3	42.4	43.5	44.6	45.7	46.8	47.9	49.0
6	28.5	29.5	30.6	31.6	32.7	33.7	34.8	35.9	36.9	38.0	39.1	40.2	41.3	42.3	43.4	44.5	45.6	46.7	47.8	48.9
7	28.4	29.4	30.5	31.6	32.6	33.7	34.7	35.8	36.9	37.9	39.0	40.1	41.2	42.3	43.3	44.4	45.5	46.6	47.7	48.8
8	28.3	29.4	30.4	31.5	32.5	33.6	34.7	35.7	36.8	37.9	38.9	40.0	41.1	42.2	43.3	44.3	45.4	46.5	47.6	48.7
9	28.3	29.3	30.4	31.4	32.5	33.5	34.6	35.7	36.7	37.8	38.9	39.9	41.0	42.1	43.2	44.3	45.3	46.4	47.5	48.6
10	28.2	29.2	30.3	31.3	32.4	33.5	34.5	35.6	36.6	37.7	38.8	39.8	40.9	42.0	43.1	44.2	45.2	46.3	47.4	48.5
11	28.1	29.2	30.2	31.3	32.3	33.4	34.4	35.5	36.6	37.6	38.7	39.8	40.8	41.9	43.0	44.1	45.2	46.2	47.3	48.4
12	28.1	29.1	30.2	31.2	32.3	33.3	34.4	35.4	36.5	37.6	38.6	39.7	40.8	41.8	42.9	44.0	45.1	46.2	47.2	48.3
13	28.0	29.0	30.1	31.1	32.2	33.2	34.3	35.4	36.4	37.5	38.5	39.6	40.7	41.7	42.8	43.9	45.0	46.1	47.1	48.2
14	27.9	29.0	30.0	31.1	32.1	33.2	34.2	35.3	36.3	37.4	38.5	39.5	40.6	41.7	42.7	43.8	44.9	46.0	47.1	48.1
15	27.9	28.9	30.0	31.0	32.1	33.1	34.2	35.2	36.3	37.3	38.4	39.4	40.5	41.6	42.7	43.7	44.8	45.9	47.0	48.0
16	27.8	28.8	29.9	30.9	32.0	33.0	34.1	35.1	36.2	37.2	38.3	39.4	40.4	41.5	42.6	43.6	44.7	45.8	46.9	48.0
17	27.7	28.8	29.8	30.9	31.9	33.0	34.0	35.1	36.1	37.2	38.2	39.3	40.4	41.4	42.5	43.6	44.6	45.7	46.8	47.9
18	27.7	28.7	29.7	30.8	31.8	32.9	33.9	35.0	36.0	37.1	38.1	39.2	40.3	41.3	42.4	43.5	44.5	45.6	46.7	47.8
19	27.6	28.6	29.7	30.7	31.8	32.8	33.9	34.9	36.0	37.0	38.1	39.1	40.2	41.2	42.3	43.4	44.4	45.5	46.6	47.7
20	27.5	28.6	29.6	30.7	31.7	32.7	33.8	34.8	35.9	36.9	38.0	39.0	40.1	41.2	42.2	43.3	44.4	45.4	46.5	47.6
21	27.5	28.5	29.5	30.6	31.6	32.7	33.7	34.8	35.8	36.9	37.9	39.0	40.0	41.1	42.1	43.2	44.3	45.3	46.4	47.5
22	27.4	28.4	29.5	30.5	31.6	32.6	33.6	34.7	35.7	36.8	37.8	38.9	39.9	41.0	42.1	43.1	44.2	45.2	46.3	47.4
23	27.3	28.4	29.4	30.5	31.5	32.5	33.6	34.6	35.7	36.7	37.8	38.8	39.9	40.9	42.0	43.0	44.1	45.2	46.2	47.3
24	27.3	28.3	29.3	30.4	31.4	32.5	33.5	34.5	35.6	36.6	37.7	38.7	39.8	40.8	41.9	42.9	44.0	45.1	46.1	47.2
25	27.2	28.2	29.3	30.3	31.4	32.4	33.4	34.5	35.5	36.6	37.6	38.7	39.7	40.8	41.8	42.9	43.9	45.0	46.0	47.1
26	27.1	28.2	29.2	30.2	31.3	32.3	33.4	34.4	35.4	36.5	37.5	38.6	39.6	40.7	41.7	42.8	43.8	44.9	46.0	47.0
27	27.1	28.1	29.1	30.2	31.2	32.3	33.3	34.3	35.4	36.4	37.5	38.5	39.5	40.6	41.6	42.7	43.7	44.8	45.9	46.9
28	27.0	28.0	29.1	30.1	31.1	32.2	33.2	34.3	35.3	36.3	37.4	38.4	39.5	40.5	41.6	42.6	43.7	44.7	45.8	46.8
29	27.0	28.0	29.0	30.0	31.1	32.1	33.1	34.2	35.2	36.3	37.3	38.3	39.4	40.4	41.5	42.5	43.6	44.6	45.7	46.7
30	26.9	27.9	28.9	30.0	31.0	32.0	33.1	34.1	35.1	36.2	37.2	38.3	39.3	40.4	41.4	42.4	43.5	44.5	45.6	46.6
31	26.8	27.9	28.9	29.9	30.9	32.0	33.0	34.0	35.1	36.1	37.1	38.2	39.2	40.3	41.3	42.4	43.4	44.5	45.5	46.6
32	26.8	27.8	28.8	29.8	30.9	31.9	32.9	34.0	35.0	36.0	37.1	38.1	39.1	40.2	41.2	42.3	43.3	44.4	45.4	46.5
33	26.7	27.7	28.7	29.8	30.8	31.8	32.9	33.9	34.9	36.0	37.0	38.0	39.1	40.1	41.1	42.2	43.2	44.3	45.3	46.4
34	26.6	27.7	28.7	29.7	30.7	31.8	32.8	33.8	34.9	35.9	36.9	38.0	39.0	40.0	41.1	42.1	43.1	44.2	45.2	46.3
35	26.6	27.6	28.6	29.6	30.7	31.7	32.7	33.8	34.8	35.8	36.8	37.9	38.9	39.9	41.0	42.0	43.1	44.1	45.1	46.2
36	26.5	27.5	28.6	29.6	30.6	31.6	32.7	33.7	34.7	35.7	36.8	37.8	38.8	39.9	40.9	41.9	43.0	44.0	45.1	46.1
37	26.4	27.5	28.5	29.5	30.5	31.6	32.6	33.6	34.6	35.7	36.7	37.7	38.8	39.8	40.8	41.9	42.9	43.9	45.0	46.0
38	26.4	27.4	28.4	29.4	30.5	31.5	32.5	33.5	34.6	35.6	36.6	37.6	38.7	39.7	40.7	41.8	42.8	43.8	44.9	45.9
39	26.3	27.3	28.4	29.4	30.4	31.4	32.4	33.5	34.5	35.5	36.5	37.6	38.6	39.6	40.7	41.7	42.7	43.8	44.8	45.8
40	26.2	27.3	28.3	29.3	30.3	31.4	32.4	33.4	34.4	35.4	36.5	37.5	38.5	39.6	40.6	41.6	42.6	43.7	44.7	45.7
41	26.2	27.2	28.2	29.2	30.3	31.3	32.3	33.3	34.3	35.4	36.4	37.4	38.4	39.5	40.5	41.5	42.6	43.6	44.6	45.7
42	26.1	27.1	28.2	29.2	30.2	31.2	32.2	33.3	34.3	35.3	36.3	37.3	38.4	39.4	40.4	41.4	42.5	43.5	44.5	45.6
43	26.1	27.1	28.1	29.1	30.1	31.1	32.2	33.2	34.2	35.2	36.2	37.3	38.3	39.3	40.3	41.4	42.4	43.4	44.4	45.5
44	26.0	27.0	28.0	29.0	30.1	31.1	32.1	33.1	34.1	35.2	36.2	37.2	38.2	39.2	40.3	41.3	42.3	43.3	44.4	45.4
45	25.9	26.9	28.0	29.0	30.0	31.0	32.0	33.0	34.1	35.1	36.1	37.1	38.1	39.2	40.2	41.2	42.2	43.2	44.3	45.3
46	25.9	26.9	27.9	28.9	29.9	30.9	32.0	33.0	34.0	35.0	36.0	37.0	38.1	39.1	40.1	41.1	42.1	43.2	44.2	45.2
47	25.8	26.8	27.8	28.8	29.9	30.9	31.9	32.9	33.9	34.9	35.9	37.0	38.0	39.0	40.0	41.0	42.1	43.1	44.1	45.1
48	25.7	26.8	27.8	28.8	29.8	30.8	31.8	32.8	33.8	34.9	35.9	36.9	37.9	38.9	39.9	41.0	42.0	43.0	44.0	45.0
49	25.7	26.7	27.7	28.7	29.7	30.7	31.7	32.8	33.8	34.8	35.8	36.8	37.8	38.8	39.9	40.9	41.9	42.9	43.9	44.9

Appendix 3 — Reduction of Observed API Gravity to API Gravity at 60°F (continued)

Observed Temperature °F	API Gravity at Observed Temperature																			
	25	26	27	28	29	30	31	32	33	34	35	36	37	38	39	40	41	42	43	44
	Corresponding API Gravity at 60°F																			
50	25.6	26.6	27.6	28.6	29.7	30.7	31.7	32.7	33.7	34.7	35.7	36.7	37.8	38.8	39.8	40.8	41.8	42.8	43.8	44.9
51	25.6	26.6	27.6	28.6	29.6	30.6	31.6	32.6	33.6	34.6	35.7	36.7	37.7	38.7	39.7	40.7	41.7	42.7	43.8	44.8
52	25.5	26.5	27.5	28.5	29.5	30.5	31.5	32.6	33.6	34.6	35.6	36.6	37.6	38.6	39.6	40.6	41.6	42.7	43.7	44.7
53	25.4	26.4	27.4	28.5	29.5	30.5	31.5	32.5	33.5	34.5	35.5	36.5	37.5	38.5	39.5	40.5	41.5	42.5	43.6	44.6
54	25.4	26.4	27.4	28.4	29.4	30.4	31.4	32.4	33.4	34.4	35.4	36.4	37.5	38.5	39.5	40.5	41.5	42.5	43.5	44.5
55	25.3	26.3	27.3	28.3	29.3	30.3	31.3	32.3	33.4	34.4	35.4	36.4	37.4	38.4	39.4	40.4	41.4	42.4	43.4	44.4
56	25.2	26.3	27.3	28.3	29.3	30.3	31.3	32.3	33.3	34.3	35.3	36.3	37.3	38.3	39.3	40.3	41.3	42.3	43.3	44.3
57	25.2	26.2	27.2	28.2	29.2	30.2	31.2	32.2	33.2	34.2	35.2	36.2	37.2	38.2	39.2	40.2	41.2	42.2	43.3	44.3
58	25.1	26.1	27.1	28.1	29.1	30.1	31.1	32.1	33.1	34.1	35.1	36.1	37.1	38.2	39.2	40.2	41.2	42.2	43.2	44.2
59	25.1	26.1	27.1	28.1	29.1	30.1	31.1	32.1	33.1	34.1	35.1	36.1	37.1	38.1	39.1	40.1	41.1	42.1	43.1	44.1
60	25.0	26.0	27.0	28.0	29.0	30.0	31.0	32.0	33.0	34.0	35.0	36.0	37.0	38.0	39.0	40.0	41.0	42.0	43.0	44.0
61	24.9	25.9	26.9	27.9	28.9	29.9	30.9	31.9	32.9	33.9	34.9	35.9	36.9	37.9	38.9	39.9	40.9	41.9	42.9	43.9
62	24.9	25.9	26.9	27.9	28.9	29.9	30.9	31.9	32.9	33.9	34.9	35.9	36.9	37.8	38.8	39.8	40.8	41.8	42.8	43.8
63	24.8	25.8	26.8	27.8	28.8	29.8	30.8	31.8	32.8	33.8	34.8	35.8	36.8	37.8	38.8	39.8	40.8	41.8	42.8	43.7
64	24.8	25.8	26.7	27.7	28.7	29.7	30.7	31.7	32.7	33.7	34.7	35.7	36.7	37.7	38.7	39.7	40.7	41.7	42.7	43.7
65	24.7	25.7	26.7	27.7	28.7	29.7	30.7	31.7	32.7	33.6	34.6	35.6	36.6	37.6	38.6	39.6	40.6	41.6	42.6	43.6
66	24.6	25.6	26.6	27.6	28.6	29.6	30.6	31.6	32.6	33.6	34.6	35.6	36.6	37.5	38.5	39.5	40.5	41.5	42.5	43.5
67	24.6	25.6	26.6	27.6	28.5	29.5	30.5	31.5	32.5	33.5	34.5	35.5	36.5	37.5	38.5	39.5	40.4	41.4	42.4	43.4
68	24.5	25.5	26.5	27.5	28.5	29.5	30.5	31.5	32.4	33.4	34.4	35.4	36.4	37.4	38.4	39.4	40.4	41.4	42.3	43.3
69	24.4	25.4	26.4	27.4	28.4	29.4	30.4	31.4	32.4	33.4	34.4	35.3	36.3	37.3	38.3	39.3	40.3	41.3	42.3	43.2
70	24.4	25.4	26.4	27.4	28.3	29.3	30.3	31.3	32.3	33.3	34.3	35.3	36.3	37.2	38.2	39.2	40.2	41.2	42.2	43.2
71	24.3	25.3	26.3	27.3	28.3	29.3	30.3	31.3	32.2	33.2	34.2	35.2	36.2	37.2	38.2	39.1	40.1	41.1	42.1	43.1
72	24.3	25.3	26.2	27.2	28.2	29.2	30.2	31.2	32.2	33.2	34.1	35.1	36.1	37.1	38.1	39.1	40.1	41.0	42.0	43.0
73	24.2	25.2	26.2	27.2	28.2	29.1	30.1	31.1	32.1	33.1	34.1	35.1	36.0	37.0	38.0	39.0	40.0	41.0	41.9	42.9
74	24.1	25.1	26.1	27.1	28.1	29.1	30.1	31.0	32.0	33.0	34.0	35.0	36.0	36.9	37.9	38.9	39.9	40.9	41.9	42.8
75	24.1	25.1	26.1	27.0	28.0	29.0	30.0	31.0	32.0	32.9	33.9	34.9	35.9	36.9	37.9	38.8	39.8	40.8	41.8	42.7
76	24.0	25.0	26.0	27.0	28.0	28.9	29.9	30.9	31.9	32.9	33.9	34.8	35.8	36.8	37.8	38.8	39.7	40.7	41.7	42.7
77	24.0	24.9	25.9	26.9	27.9	28.9	29.9	30.8	31.8	32.8	33.8	34.8	35.7	36.7	37.7	38.7	39.7	40.6	41.6	42.6
78	23.9	24.9	25.9	26.9	27.8	28.8	29.8	30.8	31.8	32.7	33.7	34.7	35.7	36.7	37.6	38.6	39.6	40.6	41.5	42.5
79	23.8	24.8	25.8	26.8	27.8	28.8	29.7	30.7	31.7	32.7	33.7	34.6	35.6	36.6	37.6	38.5	39.5	40.5	41.4	42.4
80	23.8	24.8	25.7	26.7	27.7	28.7	29.7	30.6	31.6	32.6	33.6	34.6	35.5	36.5	37.5	38.5	39.4	40.4	41.4	42.3
81	23.7	24.7	25.7	26.7	27.6	28.6	29.6	30.6	31.6	32.5	33.5	34.5	35.5	36.4	37.4	38.4	39.4	40.3	41.3	42.3
82	23.7	24.6	25.6	26.6	27.6	28.6	29.5	30.5	31.5	32.5	33.4	34.4	35.4	36.4	37.3	38.3	39.3	40.2	41.2	42.2
83	23.6	24.6	25.6	26.5	27.5	28.5	29.5	30.4	31.4	32.4	33.4	34.3	35.3	36.3	37.3	38.2	39.2	40.2	41.1	42.1
84	23.5	24.5	25.5	26.5	27.5	28.4	29.4	30.4	31.4	32.3	33.3	34.3	35.3	36.2	37.2	38.2	39.1	40.1	41.1	42.0
85	23.5	24.5	25.4	26.4	27.4	28.4	29.3	30.3	31.3	32.3	33.2	34.2	35.2	36.1	37.1	38.1	39.0	40.0	41.0	41.9
86	23.4	24.4	25.4	26.4	27.3	28.3	29.3	30.3	31.2	32.2	33.2	34.1	35.1	36.1	37.0	38.0	39.0	39.9	40.9	41.9
87	23.4	24.3	25.3	26.3	27.3	28.2	29.2	30.2	31.2	32.1	33.1	34.1	35.0	36.0	37.0	37.9	38.9	39.9	40.8	41.8
88	23.3	24.3	25.3	26.2	27.2	28.2	29.1	30.1	31.1	32.1	33.0	34.0	35.0	35.9	36.9	37.9	38.8	39.8	40.7	41.7
89	23.2	24.2	25.2	26.2	27.1	28.1	29.1	30.1	31.0	32.0	33.0	33.9	34.9	35.9	36.8	37.8	38.8	39.7	40.7	41.7
90	23.2	24.2	25.1	26.1	27.1	28.0	29.0	30.0	31.0	31.9	32.9	33.9	34.8	35.8	36.7	37.7	38.7	39.6	40.6	41.5
91	23.1	24.1	25.1	26.0	27.0	28.0	28.9	29.9	30.9	31.9	32.8	33.8	34.7	35.7	36.7	37.6	38.6	39.5	40.5	41.5
92	23.1	24.0	25.0	26.0	27.0	27.9	28.9	29.9	30.8	31.8	32.8	33.7	34.7	35.6	36.6	37.6	38.5	39.5	40.4	41.4
93	23.0	24.0	25.0	25.9	26.9	27.9	28.8	29.8	30.8	31.7	32.7	33.7	34.6	35.6	36.5	37.5	38.4	39.4	40.3	41.3
94	22.9	23.9	24.9	25.9	26.8	27.8	28.8	29.7	30.7	31.7	32.6	33.6	34.5	35.5	36.5	37.4	38.4	39.3	40.3	41.2
95	22.9	23.9	24.8	25.8	26.8	27.7	28.7	29.7	30.6	31.6	32.5	33.5	34.5	35.4	36.4	37.3	38.3	39.2	40.2	41.1
96	22.8	23.8	24.8	25.7	26.7	27.7	28.6	29.6	30.6	31.5	32.5	33.4	34.4	35.3	36.3	37.3	38.2	39.2	40.1	41.1
97	22.8	23.7	24.7	25.7	26.6	27.6	28.6	29.5	30.5	31.5	32.4	33.4	34.3	35.3	36.2	37.2	38.1	39.1	40.0	41.0
98	22.7	23.7	24.7	25.6	26.6	27.5	28.5	29.5	30.4	31.4	32.3	33.3	34.3	35.2	36.2	37.1	38.1	39.0	40.0	40.9
99	22.7	23.6	24.6	25.6	26.5	27.5	28.4	29.4	30.4	31.3	32.3	33.2	34.2	35.1	36.1	37.0	38.0	38.9	39.9	40.8

Appendices 283

Appendix 3 — Reduction of Observed API Gravity to API Gravity at 60°F (continued)

Observed Temperature °F	API Gravity at Observed Temperature																				
	25	26	27	28	29	30	31	32	33	34	35	36	37	38	39	40	41	42	43	44
	Corresponding API Gravity at 60°F																				
100	22.6	23.6	24.5	25.5	26.5	27.4	28.4	29.3	30.3	31.3	32.2	33.2	34.1	35.1	36.0	37.0	37.9	38.9	39.8	40.7
101	22.5	23.5	24.5	25.4	26.4	27.4	28.3	29.3	30.2	31.2	32.1	33.1	34.0	35.0	35.9	36.9	37.8	38.8	39.7	40.7
102	22.5	23.4	24.4	25.4	26.3	27.3	28.3	29.2	30.2	31.1	32.1	33.0	34.0	34.9	35.9	36.8	37.8	38.7	39.7	40.6
103	22.4	23.4	24.3	25.3	26.3	27.2	28.2	29.1	30.1	31.1	32.0	33.0	33.9	34.9	35.8	36.7	37.7	38.6	39.6	40.5
104	22.4	23.3	24.3	25.3	26.2	27.2	28.1	29.1	30.0	31.0	31.9	32.9	33.8	34.8	35.7	36.7	37.6	38.6	39.5	40.4
105	22.3	23.3	24.2	25.2	26.1	27.1	28.1	29.0	30.0	30.9	31.9	32.8	33.8	34.7	35.7	36.6	37.5	38.5	39.4	40.4
106	22.2	23.2	24.2	25.1	26.1	27.0	28.0	29.0	29.9	30.9	31.8	32.8	33.7	34.6	35.6	36.5	37.5	38.4	39.4	40.3
107	22.2	23.1	24.1	25.1	26.0	27.0	27.9	28.9	29.8	30.8	31.7	32.7	33.6	34.6	35.5	36.5	37.4	38.3	39.3	40.2
108	22.1	23.1	24.0	25.0	26.0	26.9	27.9	28.8	29.8	30.7	31.7	32.6	33.6	34.5	35.4	36.4	37.3	38.3	39.2	40.1
109	22.1	23.0	24.0	24.9	25.9	26.9	27.8	28.8	29.7	30.7	31.6	32.5	33.5	34.4	35.4	36.3	37.3	38.2	39.1	40.1
110	22.0	23.0	23.9	24.9	25.8	26.8	27.7	28.7	29.6	30.6	31.5	32.5	33.4	34.4	35.3	36.2	37.2	38.1	39.0	40.0
111	22.0	22.9	23.9	24.8	25.8	26.7	27.7	28.6	29.6	30.5	31.5	32.4	33.4	34.3	35.2	36.2	37.1	38.0	39.0	39.9
112	21.9	22.9	23.8	24.8	25.7	26.7	27.6	28.6	29.5	30.5	31.4	32.3	33.3	34.2	35.2	36.1	37.0	38.0	38.9	39.8
113	21.8	22.8	23.8	24.7	25.7	26.6	27.6	28.5	29.5	30.4	31.3	32.3	33.2	34.2	35.1	36.0	37.0	37.9	38.8	39.7
114	21.8	22.7	23.7	24.6	25.6	26.5	27.5	28.4	29.4	30.3	31.3	32.2	33.2	34.1	35.0	36.0	36.9	37.8	38.7	39.7
115	21.7	22.7	23.6	24.6	25.5	26.5	27.4	28.4	29.3	30.3	31.2	32.1	33.1	34.0	35.0	35.9	36.8	37.7	38.7	39.6
116	21.7	22.6	23.6	24.5	25.5	26.4	27.4	28.3	29.3	30.2	31.1	32.1	33.0	34.0	34.9	35.8	36.7	37.7	38.6	39.5
117	21.6	22.6	23.5	24.5	25.4	26.4	27.3	28.3	29.2	30.1	31.1	32.0	32.9	33.9	34.8	35.7	36.7	37.6	38.5	39.4
118	21.6	22.5	23.5	24.4	25.4	26.3	27.2	28.2	29.1	30.1	31.0	31.9	32.9	33.8	34.7	35.7	36.6	37.5	38.5	39.4
119	21.5	22.4	23.4	24.3	25.3	26.2	27.2	28.1	29.1	30.0	30.9	31.9	32.8	33.7	34.7	35.6	36.5	37.5	38.4	39.3
120	21.4	22.4	23.3	24.3	25.2	26.2	27.1	28.1	29.0	29.9	30.9	31.8	32.7	33.7	34.6	35.5	36.5	37.4	38.3	39.2
121	21.4	22.3	23.3	24.2	25.2	26.1	27.1	28.0	28.9	29.9	30.8	31.7	32.7	33.6	34.5	35.5	36.4	37.3	38.2	39.1
122	21.3	22.3	23.2	24.2	25.1	26.1	27.0	27.9	28.9	29.8	30.8	31.7	32.6	33.5	34.5	35.4	36.3	37.2	38.2	39.1
123	21.3	22.2	23.2	24.1	25.1	26.0	26.9	27.9	28.8	29.8	30.7	31.6	32.5	33.5	34.4	35.3	36.2	37.2	38.1	39.0
124	21.2	22.2	23.1	24.1	25.0	25.9	26.9	27.8	28.8	29.7	30.6	31.5	32.5	33.4	34.3	35.3	36.2	37.1	38.0	38.9
125	21.2	22.1	23.0	24.0	24.9	25.9	26.8	27.8	28.7	29.6	30.6	31.5	32.4	33.3	34.3	35.2	36.1	37.0	37.9	38.9
126	21.1	22.0	23.0	23.9	24.9	25.8	26.8	27.7	28.6	29.6	30.5	31.4	32.3	33.3	34.2	35.1	36.0	36.9	37.9	38.8
127	21.0	22.0	22.9	23.9	24.8	25.8	26.7	27.6	28.6	29.5	30.4	31.4	32.3	33.2	34.1	35.0	36.0	36.9	37.8	38.7
128	21.0	21.9	22.9	23.8	24.8	25.7	26.6	27.6	28.5	29.4	30.4	31.3	32.2	33.1	34.1	35.0	36.0	36.8	37.7	38.6
129	20.9	21.9	22.8	23.8	24.7	25.6	26.6	27.5	28.4	29.4	30.3	31.2	32.1	33.1	34.0	34.9	35.8	36.7	37.6	38.6
130	20.9	21.8	22.8	23.7	24.6	25.6	26.5	27.4	28.4	29.3	30.2	31.2	32.1	33.0	33.9	34.8	35.8	36.7	37.6	38.5
131	20.8	21.8	22.7	23.6	24.6	25.5	26.5	27.4	28.3	29.2	30.2	31.1	32.0	32.9	33.8	34.8	35.7	36.6	37.5	38.4
132	20.8	21.7	22.6	23.6	24.5	25.5	26.4	27.3	28.2	29.2	30.1	31.0	31.9	32.9	33.8	34.7	35.6	36.5	37.4	38.3
133	20.7	21.6	22.6	23.5	24.5	25.4	26.3	27.3	28.2	29.1	30.0	31.0	31.9	32.8	33.7	34.6	35.5	36.5	37.4	38.3
134	20.6	21.6	22.5	23.5	24.4	25.3	26.3	27.2	28.1	29.0	30.0	30.9	31.8	32.7	33.6	34.6	35.5	36.4	37.3	38.2
135	20.6	21.5	22.5	23.4	24.3	25.3	26.2	27.1	28.1	29.0	29.9	30.8	31.7	32.7	33.6	34.5	35.4	36.3	37.2	38.1
136	20.5	21.5	22.4	23.3	24.3	25.2	26.1	27.1	28.0	28.9	29.8	30.8	31.7	32.6	33.5	34.4	35.3	36.2	37.1	38.0
137	20.5	21.4	22.4	23.3	24.2	25.2	26.1	27.0	27.9	28.9	29.8	30.7	31.6	32.5	33.4	34.4	35.3	36.2	37.1	38.0
138	20.4	21.4	22.3	23.2	24.2	25.1	26.0	27.0	27.9	28.8	29.7	30.6	31.6	32.5	33.4	34.3	35.2	36.1	37.0	37.9
139	20.4	21.3	22.2	23.2	24.1	25.0	26.0	26.9	27.8	28.7	29.7	30.6	31.5	32.4	33.3	34.2	35.1	36.0	36.9	37.8

Appendix 4. Galvanic Series of Metals and Metal Alloys

ANODIC *Corroded End*

Magnesium
Zinc
Galvanized Steel
Aluminum
Mild Steel
Cast Iron
50-50 Tin-Lead Solder
Stainless Steel Type 304 (active)
Stainless Steel Type 316 (active)
Lead
Tin
Muntz Metal
Inconel
Hastelloy
Yellow Brass
Admiralty Brass
Aluminum Bronze
Red Brass
Copper
Monel
Stainless Steel Type 304 (passive)
Stainless Steel Type 316 (passive)
Silver
Graphite
Gold
Platinum

CATHODIC *(protected end)*

Appendix 5. Composition of Synthetic Sea Water Utilized in ASTM D-665-B

Salt	g/L
NaCl	24.54
$MgCl_2 \cdot 6H_2O$	11.10
Na_2SO_4	4.09
$CaCl_2$	1.16
KCl	0.69
$NaHCO_3$	0.20
KBr	0.10
H_3BO_3	0.03
$SrCl_2 \cdot 6H_2O$	0.04
NaF	0.003

Appendix 6. Factors and Metal Densities Needed to Obtain ipy and mdd Corrosion Rates

To obtain mdd

ipy (696 × density) = mdd

To obtain ipy

mdd (0.00144 ÷ density) = ipy
mpy × 1000 = ipy

Metal	Density, g/cm^3
Aluminum	2.72
Brass (red)	8.75
Brass (yellow)	8.47
Copper	8.92
Copper-nickel (70-30)	8.95
Iron	7.87
Lead	11.35
Magnesium	1.74
Nickel	8.89
Monel	8.84
Silver	10.50
Titanium	4.54
Tin	7.29
Zinc	7.14

ipy = inches per year
mpy = mils per year
mdd = milligrams per square decimeter per day

Appendix 7. Compatibility of Various Materials with Common Fuels and Solvents

Elastomer or Metal	Diesel Fuel	Unleaded Gasoline	Jet Fuel JP-4, JP-5	Kerosene	Mineral Spirits	Heavy Fuel Oil	Xylene
			Hydrocarbon				
Ryton	OK	OK	OK	OK	OK	OK	OK
Polycarbonate	OK	OK	POOR	POOR	MARGINAL	OK TO MARGINAL	POOR
Polypropylene	OK	MARGINAL	OK TO MARGINAL	OK TO MARGINAL	OK TO MARGINAL	OK	OK TO MARGINAL
PTFE (Teflon)	OK	OK	OK	OK	OK	OK TO MARGINAL	OK
PVC	OK	MARGINAL	OK	OK	OK	OK	POOR
Natural Rubber	POOR	POOR	POOR	POOR	POOR	POOR	POOR
Buna-N (Nitrile)	OK	OK	OK	OK	OK	POOR	POOR
Neoprene	OK TO MARGINAL	OK TO MARGINAL	OK	OK	MARGINAL	OK TO MARGINAL	POOR
Silicone	POOR	POOR	POOR	POOR	POOR	POOR	POOR
Tygon	NO INFO.	MARGINAL	POOR	POOR	OK TO MARGINAL	OK	POOR
Viton	OK	OK	OK	OK	OK	OK	POOR
304 Stainless	OK	OK	OK	OK	OK	OK	OK TO MARGINAL
316 Stainless	OK	OK	OK	OK	OK	OK	OK TO MARGINAL
Aluminum	OK	OK	OK	OK	OK	MARGINAL	OK
Copper	OK	OK TO MARGINAL	OK TO MARGINAL	OK TO MARGINAL	NO INFO.	OK	OK
Brass	OK	NO INFO.	OK	OK	NO INFO.	OK TO MARGINAL	OK
Cast Iron	OK	OK	OK	OK	OK TO MARGINAL	OK	OK TO MARGINAL
Hastelloy	OK TO MARGINAL	OK	OK TO MARGINAL	OK TO MARGINAL	OK TO MARGINAL	OK	OK
Titanium	OK TO MARGINAL	OK	OK	OK	OK TO MARGINAL	OK	OK
Carbon Graphite	OK	OK	OK	OK	OK	OK	OK

Useful Terms and Definitions

A variety of technical terms are used throughout this manuscript. Most of these terms are familiar and known by those who have worked in the petroleum industry for a few years. The purpose of providing this glossary is to focus on the terminology and definitions which are the most useful when describing fuel performance problems.

Absolute Pressure   A pressure scale with the baseline zero point at perfect vacuum.

Accumulator   A vessel containing fluid stored under pressure which acts to serve as a source of fluid power.

Actuator   A mechanism used to convert hydraulic energy into mechanical energy.

Additive   A chemical substance added to a product to impart or improve certain properties. Typical fuel additives include antioxidants, cetane improvers, corrosion inhibitors, demulsifiers, detergents, dyes, metal deactivators, octane improvers and wax crystal modifiers.

Alkylation   A refinery process for producing high-octane components consisting mainly of branched chain paraffins. The process involves combining light olefins with isoparaffins, usually butene and isobutane, in the presence of a strong acid catalyst such as hydrofluoric or sulfuric acid.

Aluminum Chloride Processing   A refining method using aluminum chloride as a catalyst to improve the appearance and odor of steam cracked naphtha streams. Aluminum chloride functions as a catalyst for the polymerization of olefins into higher molecular weight, less problematic compounds.

Anode   The electrode of an electrolytic cell where oxidation occurs. Electrons move away from the anode in an external circuit. At the anode, corrosion occurs and metal ions are solubilized.

Antifoam Agent   An additive, usually silicone or polyglycol based, used to reduce foaming in petroleum products.

API (American Petroleum Institute)   A trade association of firms engaged in all aspects of the U.S. petroleum industry including producers, refiners, marketers and transporters.

Aromatics   Unsaturated hydrocarbons such as benzene, toluene, xylene and related compounds. Aromatic compounds share electrons equally among carbon atoms.

Ash Content In the petroleum industry, this term usually refers to metals and metal salts contained in crude oil or a petroleum product. These noncombustible substances can form deposits and can impair engine efficiency and power.

Asphaltenes These are complex high molecular weight polycyclic aromatic compounds which may contain oxygen, sulfur or nitrogen heteroatoms. They are found in crude oil and in certain heavy fuel oils in micellar form. Their dispersion throughout oil can be stabilized by asphaltene precursors called resins and maltenes.

Paraffins can agglomerate and crystallize onto asphaltenes. Under high shear or high temperature conditions these asphaltenes will tend to coalesce and deposit onto pump parts and other moving components. Also, if blended with other oils, asphaltenes may flocculate and become insoluble in the mixture.

Some of the fuel performance problems which can be associated with asphaltenes include:
- Pump deposit buildup
- Restricted flow in lines and piping
- Tank bottom sludge and deposit accumulation
- Stabilization of existing fuel - water emulsions

Autoignition The spontaneous ignition of a mixture of fuel and air without an ignition source. Diesel and jet engines combust fuel in this manner.

Ball-On-Three-Disks (BOTD) This test method is utilized to measure the lubricity characteristics of distillate fuel. This method involves rotating a ball against three disks lubricated by distillate fuel. After testing, the wear scar diameters on three disks are measured and averaged. Wear scar diameters of approximately 0.45 mm or less are presently considered acceptable.

Bar A value equal to 0.987 atmospheres or 29.53 inches of mercury.

Barrel A unit of liquid volume utilized by the petroleum industry equal to 42 U.S. gallons or approximately 35 Canadian gallons.

Blow-by Combustion gases which are forced past the piston rings and into the crankcase of an internal combustion engine. Blow-by gases may contain NO_x, SO_x, CO, CO_2, water vapor, hydrocarbon gases and soot. In combination with water, some of these gases can form corrosive acids.

Borderline Pumping Temperature The lowest temperature at which a lubricating oil can be efficiently pumped. It is also identified as the temperature at which a lubricant's viscosity equals 30,000 cPs.

Bosch Number A measurement of diesel exhaust smoke color: clear = 0; black = 10.

Bottoms The high boiling residual liquid from the refining process; also called residuum. Any residue remaining in a distillation unit after the highest boiling material has been removed by distillation. The boiling range will depend upon the feedstock and the amount of material removed.

Brake Specific Fuel Consumption (BSFC) This is a measure of the pounds of fuel consumed per brake horsepower for a one hour period.

Brake Specific Horsepower (BSHP) This is a measure of the power output at the end of the crankshaft available for doing work. 1.36 BHP = 1.00 kW.

Bright Stock High viscosity oil which is highly refined and dewaxed and produced from residual lubestock or bottoms. It is used in finished lubricant blends to provide good bearing film strength, prevent scuffing and reduce oil consumption.

British Thermal Unit (BTU) A measure of the quantity of heat required to raise the temperature of one pound of water 1°F. For fuels such as gasoline, kerosene, fuel oil and residual fuels the following formulas can be used to determine the BTU per pound:

Sherman & Kropff Equation:
For gasoline:	$18{,}320 + 40(°API - 10)$
For kerosene:	$18{,}440 + 40(°API - 10)$
For fuel oil:	$18{,}650 + 40(°API - 10)$

Faragher, Morrell & Essex Equation:
For heavy residual fuels:	$17{,}645 + (54 \times °API)$
For Bunker "C" fuel:	$17{,}685 + (57.9 \times °API)$

These equations can be used to calculate the gross heating value or higher heating value of fuel whereby water formed during combustion is condensed and combustion gases are cooled to the initial fuel-air temperature.

Brookfield Viscosity A measurement of the apparent viscosity of a fluid using a Brookfield viscometer. This viscometer is a variable sheer, variable RPM viscometer.

BS&W "Bottoms, Sediment and Water." This centrifuge method is used to measure the approximate amount of suspended solids and water in crude oil and petroleum products.

Buna-N This nitrile rubber material has been used extensively in automatic transmissions. It is compatible with most petroleum products, can be formed into a variety of shapes and is economical. Useful temperature ranges from -40°F to 230°F (-40°C to 110°C). It has excellent resistance to swelling and softening at higher temperatures.

Camshaft A lobed shaft usually designed to rotate in relation to the speed of an engine. The lobes on this shaft are somewhat elliptical in shape. When rotating, the lobes act to push against a rod or some other mechanism to operate other parts of the valve train.

Carbon Black Oil Usually a viscous, highly aromatic residual oil utilized in the manufacture of carbon black. These oils may also contain polynuclear aromatic compounds, hydrogen sulfide and ash.

Cathode The portion of a corrosion cell in which reduction occurs. This region is not visibly affected.

Cavitation A condition which exists when gas accumulates within a liquid stream being pumped. This area of low pressure can lead to pump wear and damage.

Centipoise A standard unit of viscosity equal to 0.01 poise. At 20°C, water has a viscosity or 1.002 centipoise or 0.01002 poise.

Cetane Index A calculated value which incorporates fuel distillation properties and °API which can be used to approximate the cetane number of a fuel. This method, however, does not provide information on the performance provided by a cetane improver.

The cetane index of distillate fuel can be related to the aromatic content of the blend. As fuel aromatic content increases, the cetane index will typically decrease. A general relationship exists which relates a cetane index of 40 with a diesel fuel aromatic content of about 35%.

Cetane Number A fuel performance measurement utilizing a standard engine and standard reference fuels. The property of fuel ignition delay is determined by this engine. High cetane fuels have a shorter ignition delay period. High cetane fuels provide for easy starting, quiet running and more complete combustion.

This number is determined from the percentage of n-cetane which must be mixed with heptamethylnonane to give the same ignition performance as the fuel being tested. The cetane number can be obtained by using the following calculation:

$$\text{Cetane Number} = \% \; n\text{-cetane} + 0.15(\% \; \text{heptamethylnonane})$$

CFPP (Cold Filter Plugging Point) A measure of the ability of a distillate fuel to be filtered satisfactorily in cold environments. This test measures the temperature at which fuel wax crystals can reduce or halt the flow of fuel through a standardized test filter.

Channeling A term used to describe lubricants, especially gear oils, which fail to readily flow back together when a machine component passes through a volume of cold oil. When channeling occurs, an oil is unable to flow and lubricate.

Check Valve A valve which permits flow of fluid in only one direction. Often a check valve can be nothing more than a spring, ball and seat separating two openings. Common valve types include in-line check valves and right angle check valves.

Claus Sulfur Recovery Process The Claus process is a controlled combustion process commonly used for the recovery of sulfur from H_2S. Temperatures >2000°F are achieved during combustion and yields of about 95% are typical. The basic reaction involves the following:

$$2H_2S + 3O_2 \rightarrow 2SO_2 + 2H_2O$$
$$2SO_2 + 4H_2S \rightarrow S_6 + 4H_2O$$

Clay Treating A process used to improve the color of cracked naphthas and light distillates. It is also used to remove surface active agents which can negatively impact the WSIM rating of jet fuel.

Cloud Point This is the temperature at which a "cloud" or haze of wax crystals appears when a fuel or lubricant is cooled under standard test conditions.

Coefficient of Friction The coefficient of friction between two surfaces is the ratio of the force required to move one surface over the other to the total force pressing the two together.

Coking A refinery process in which fuel oil is converted to lighter boiling liquids and coke by a thermal cracking process.

Cold Cranking Simulator (CCS) An intermediate shear rate viscometer that predicts the ability of an oil to permit a satisfactory cranking speed to be developed in a cold engine.

Compression Ignition A fuel combustion process whereby heat generated by the rapid compression of air is utilized as an autoignition source for fuel. This form of fuel combustion is utilized in the diesel engine. No spark plug is required to ignite the fuel.

Compression Ratio In an internal combustion engine it is the ratio of the volume of the combustion space at the bottom of the piston stroke (bottom dead center) to that at the top of the piston stroke (top dead center). High compression ratios usually improve fuel economy but increase NO_x emissions. Compression ratio values for spark-ignition engines typically range from 7:1 to 9:1 to as high as 12:1 for higher RPM engines. For diesel engines, compression ratios typically range from 14:1 to 18:1 to as high as 24:1 for lighter, higher RPM engines.

Conductivity The property of a substance which allows the passage of electrical charge. Fuels typically are poor conductors of electrical charge and enable static charge to build. Often, conductivity aids are added to fuel to permit the transfer and release of static charge.

Conjugation In fuel technology, this term typically refers to olefins containing double bonds which exist between alternating carbon atoms. Conjugated olefins are quite unstable and lead to fuel degradation and deposits.

$$CH_2=CH-CH=CH-CH=CH-R$$
A Conjugated Olefin

Conradson Carbon Number ASTM D-189 Determination of the weight of non-volatile residue formed after evaporation and atmospheric pyrolysis of fuel or oil. This test method provides some information about the relative coke-forming or deposit forming tendency of a fuel or oil. Products having a high ash value will have an erroneously high carbon residue value.

Continuous Spray Pump Fuel injection pump which provides a constant rather than an intermittent spray; normally for fuel injection in spark ignition engines.

Cracking A refining process in which high molecular weight liquid hydrocarbons are broken down into lighter, lower boiling liquids or gases. This process can be accomplished thermally or catalytically.

Crankcase Dilution Dilution of lubricating oil in the oil pan or sump of an engine by fuel which has entered the crankcase.

Degree-days The difference between the daily mean temperature and 65°F when the temperature is below 65°F for a given day. Degree-day fuel oil delivery systems are based on the principle that the consumption of fuel oil for heating is roughly proportionate to the number of degree days. The number of degree-days for a given period is arrived at by adding together the number of degree-days for individual days in the period.

Demulsibility The ability of a fuel or oil to separate from water.

Detergent In relationship to fuel technology, a detergent is an oil soluble surfactant added to fuel aiding in the prevention and removal of deposits. Examples include **anionic** alkyl aryl sulfonates, **cationic** fatty acid amides or **non-ionic** polyol condensates.

Dezincification Some brass alloys are susceptible to pitting corrosion or loss of zinc from the metal matrix. This type of corrosion usually occurs when metal is in contact with high percentages of oxygen and carbon dioxide.

Dimer Acid Produced by the reaction and combination of two unsaturated fatty acids at mid-molecule to form a single molecule.

Direct Injection A type of fuel injection process used in larger commercial and stationary diesel engines. In this process, fuel is injected at high pressure, through a multiple-hole nozzle directly into the combustion chamber. Combustion chamber design, injection pressure and air flow are optimized to maximize combustion and fuel efficiency. Direct injected engines are typically more fuel efficient than indirect injected engines, but usually generate more noise.

Displacement The volume of fluid discharged by a pump per revolution or cycle minus loss due to leakage. Terms gallons/minute or liters/minute are commonly used to describe the **delivery** of fluid by the pump. To calculate delivery, the following conversion factors can be used:

$$\text{Delivery (gal/min)} = \frac{\text{displacement (in}^3) \times \text{RPM}}{\text{Number of revolutions}} \div 231$$

$$\text{Delivery (liters/min)} = \frac{\text{displacement (ml)} \times \text{RPM}}{\text{Number of Revolutions}} \div 1000$$

Doctor Test A method which can be used to determine the *sweet or sour* nature of gasoline by treating gasoline with a solution of lead plumbite and sulfur. As an example, the reaction which occurs with methylmercaptan is as follows:

$$2CH_3SH + Na_2PbO_2 \rightarrow CH_3\text{-}S\text{-}Pb\text{-}S\text{-}CH_3 + 2NaOH$$

The addition of sulfur results in the formation of black lead sulfide:

$$(CH_3\text{-}S)_2Pb + S \rightarrow CH_3\text{-}S\text{-}S\text{-}CH_3 + PbS$$

and

$$2CH_3SH + Na_2PbO_2 + S \rightarrow CH_3\text{-}S\text{-}S\text{-}CH_3 + PbS + 2NaOH$$

When used in fuel processing, the PbS and NaOH can be reacted with air and heat. This process results in reformation of the original Doctor Solution:

$$PbS + 4NaOH + 2O_2 \rightarrow Na_2PbO_2 + Na_2SO_4 + 2H_2O$$

Drivability Index This value is calculated by utilizing parameters from the ASTM D-86 distillation profile. The purpose of this measurement is to be able to predict the cold-start drivability of gasoline. It is calculated as follows:

$$DI = 1.5(T_{10}) + 3.0(T_{50}) + (T_{90})$$

where:
T_{10} is the 10% evaporation temperature
T_{50} is the 50% evaporation temperature
T_{90} is the 90% evaporation temperature

Lower drivability index values reflect better cold-start drivability.

Emulsion A dispersed, two-phase system in which one phase is usually water and the other oil. An emulsion in which oil is dispersed in water is termed an oil-in-water emulsion. An emulsion in which water is dispersed in oil is termed a water-in-oil emulsion.

In a dispersed system, it is possible to have both phases in existence at the same time. However, whenever fuels emulsify with water, water-in-oil emulsion typically form. Agents which comprise the external phase of an emulsion are usually the most soluble in the bulk liquid in which the emulsion exists.

Endothermic Reaction A reaction in which heat is absorbed. To maintain a constant temperature during reaction, heat must be added to reactants and products.

Exhaust Gas Recirculation (EGR) The mixing of exhaust gas with intake air used in fuel combustion. In the diesel engine, mixing of exhaust gas helps in reducing the NO_x emissions. This process lowers the combustion temperature and oxygen concentration, thus lowering the total NO_x. However, excessive EGR leads to the formation of increasing amounts of CO, soot and hydrocarbon emissions.

Exothermic Reaction A reaction which evolves heat. Refining processes such as alkylation, hydrogenation and polymerization are exothermic.

Ferric Hydroxide A fragile, reddish brown product of iron or steel corrosion, $Fe(OH)_3$.

Ferrous Hydroxide A product of iron corrosion which is white in appearance, $Fe(OH)_2$. When ferric ions enter, the product will appear black, brown or green in color.

Ferrous Oxide (Hydrous) A hydrated, jet black product of iron corrosion, $FeO \bullet nH_2O$.

Fifteen/Five (15/5) Distillation A laboratory distillation which is also referred to as the "True Boiling Point" distillation. It is performed on a column containing 15 theoretical plates at a reflux ratio of 5:1.

Fines Typically broken or crushed catalyst support media composed of alumina and silica. Smaller fines are approximately 20 microns in diameter.

Flame Front The region between the burning and unburned zone present in gas phase combustion processes or the leading edge of burning fuel within the combustion chamber. In an internal combustion engine, movement of the flame front within the confines of the combustion chamber can influence overall quality of the combustion process.

Good combustion is characterized by a flame burning uniformly across the piston head. Deposits and poor quality fuel can interfere with the movement of the flame across the combustion chamber. Rough running can result.

Flash Point The lowest temperature at which the vapor above a liquid will ignite when exposed to a flame. The flash point is an important indicator of the fire and explosion hazards associated with a petroleum product.

Floating Roof A special type of storage tank roof which floats on the surface of the oil in storage. This helps to eliminate tank venting and reduces evaporative loss.

Flooding A condition that prevents starting of an engine when more fuel is drawn in than can be ignited.

Fluidizer Oil Compounds such as polyisobutylene or refined naphthenic oils which aid in the removal of deposits from the underside of intake valves.

Fuel Oil A general term applied to petroleum products used for the production of power or heat with a flash point >100°F. Fuel oils are typically classified as either engine fuels or burner fuels.

A classification system has been developed to describe various fuel oil types. For example #1 fuel oil is similar to kerosene, #2 fuel oil is similar to diesel fuel and #4 fuel oil is viscous oil at room temperature and is typically used to fuel industrial furnaces. The #5 and #6 fuel oil classifications describe viscous oils which must be heated before burning. These oils are used as bunker fuels in ships and industrial power plants.

Fuel Oil Equivalent (FOE) The heating value of a standard barrel of fuel oil, equal to 6.05×10^6 BTU.

Fungible Fuel which is available from different sources but can be freely exchanged and used as a single source fuel. No interactions will occur between the fuels.

Furfural A colorless liquid which changes to reddish brown upon exposure to light and air. Furfural forms condensation products with many types of compounds including phenols and amines. If present in kerosene or jet fuel, furfural may lead to color degradation.

Galvanic Cell A cell in which a chemical change is the source of electrical energy. The cell is established when two dissimilar metals are in contact or two similar metals in different electrolytic solutions are in contact.

Gas Oil A refinery fraction boiling within a typical temperature range between 330°F and 750°F. Diesel fuel, heating oil, kerosene and heating oil fall into this fraction.

Glow Plug A device used to provide heat to the prechamber of an indirect injected diesel engine. It is used in cold conditions to permit ease of starting. The glow plugs have a preheating time of about 7 seconds.

Gum A term used to describe dark colored deposits which form upon the oxidation of compounds present in fuel. This term has also been used to describe the products of poor combustion deposited within engine cylinders.

Heavy Fuel Oil A residue of crude oil refining. The product remaining after gasoline, kerosene, diesel fuel, lubricating oil and other distillates have been removed.

Heterocycle A compound which contains a closed ring system in which the atoms are of more than one kind. Sulfur, nitrogen and oxygen are common heteroatoms found in petroleum heterocycles.

Horsepower The power required to lift 550 pounds one foot in height in one second or 33,000 pounds one foot in one minute. One horsepower is equal to 746 watts or 42.4 BTU/min.

Hydrodynamics The science which deals with liquids in motion and the associated kinetic energy.

Hydrophilic A substance which has an affinity for water. In fuel technology, certain polar organic compounds such as alcohols, organic acids and amines are hydrophilic.

Hydrophobic A substance which lacks affinity for water. In fuel technology, hydrocarbon species are hydrophobic.

Indirect Injection A type of diesel fuel injection process used primarily in smaller engines to achieve good mixing of injected fuel into air. Fuel is not injected directly into the combustion chamber, but into a smaller, pre-chamber which feeds into the primary combustion chamber. When compressed air enters the pre-chamber, fuel is sprayed from the fuel injector at a relatively low pressure to begin the combustion process. The pressure of the burning fuel forces the mixture into the primary combustion chamber where thorough mixing with air occurs. Completion of the combustion process takes place at a lower pressure and at a relatively lower noise level than most direct injected engines. Indirect injected engines are often fitted with a glow plug to aid in cold starting of the engine.

Induction Time The time period under given test conditions in which a petroleum product does not absorb oxygen at a substantial rate to form gum. It is measured as the time elapsed between the placing of a test bomb containing 50 ml of fuel and oxygen at 100-102 psi into a 100°C test well and the point in the time-pressure curve preceded by two consecutive 15-minute periods of a pressure drop not less than 2 psi.

Isomerization A refinery process used to produce branched compounds from straight-chain molecules. Butane, pentane and hexane are isomerized to branched, higher octane compounds through isomerization. Isomerization reactions are catalyzed through the use of aluminum chloride and other metal catalysts.

JFTOT An acronym for Jet Fuel Thermal Oxidation Test. It is a measurement by ASTM D-3241 to determine the tendency of jet fuel to form deposits when heated.

Kauri-Butanol (KB) Value A measure of the relative solvent power of a hydrocarbon; the higher the KB value, the greater the effect of the hydrocarbon as a solvent.

Kinematic Viscosity A coefficient defined as the ratio of the dynamic viscosity of a fluid to its density. The *centistoke* is the reported value of kinematic viscosity measurement.

Laminar Flow A condition where fluid mass moves in a continuous, streamlined parallel path.

Light-Ends A term used to describe the low boiling point fractions in a fuel or refined petroleum fraction.

Liquefied Petroleum Gas (LPG) Compressed and liquefied light-end petroleum fractions. LPG is primarily propane with low concentrations of ethane and butane.

Manifold A common conducting unit for fluids and gases into which multiple ports are connected.

Merox Sweetening A process used in fuel refining to remove primarily hydrogen sulfide and mercaptans from fuel. This is a patented, proprietary process.

Mercaptan A thiol compound containing the structure R-SH where R is carbon. These compounds are liquids and possess strong, unpleasant odors. They are quite common in crude oil and can be found in finished fuels.

Meter-In To regulate the amount of flow into a system.

Meter-Out To regulate the amount of flow out of a system.

mpy Weight Loss An acronym for mils per year weight loss. This rating is used to describe the loss in the thickness of a metal surface often due to corrosion.

Naphtha A term used to describe a light crude oil distillate fraction boiling in the gasoline range. The typical boiling range is from about 100°F to 420°F. Naphthas can be subdivided into the following fractions:

IBP	-	160°F	Light naphtha
160°F	-	280°F	Intermediate naphtha
280°F	-	330°F	Heavy naphtha
330°F	-	420°F	High boiling naphtha

Naphthenes Saturated hydrocarbons found in fuels which contain at least one closed ring system.

Naphthenic Acids   Acids derived from crude oil which are usually monocarboxylic, monocyclic and completely saturated. Many are derivatives of cyclopentane and more complex alicyclic ring systems.

These acids can also be produced by oxidation of cycloparaffins during distillation or other refining processes. The boiling range of these compounds is usually between about 400°F to 575°F with molecular weights ranging from 180 to 350. They are frequently removed from petroleum product by caustic washing. The sodium salts of these acids are water soluble and act as detergents and emulsifying agents.

Natural Gasoline   Hydrocarbons condensed from natural gas which consist primarily of pentanes and heavier molecular weight compounds.

Neoprene   Neoprene is compatible with most petroleum products and is used in the manufacture of seals and gaskets. However, above temperatures of 150°F (66°C), neoprene becomes unstable and less useful.

Neutralization Number   The amount of acid or base required to neutralize all components present.

Newtonian Fluid   A fluid with a constant viscosity at a given temperature regardless of the shear rate.

Octane   A hydrocarbon with the formula C_8H_{18} having 18 different isomers. These isomers have typical boiling points between 99°C and 125°C. The most important of these isomers is 2,2,4-trimethyl pentane or isooctane. It is a colorless liquid with a boiling point of 99°C and is used as a standard for rating the antiknock properties of gasoline.

O-Ring   An O-ring is a seal or gasket formed in the shape of a circle or letter "O." It can be made from a variety of elastomeric materials and can be is used to provide a very effective seal in high pressure static and dynamic systems. It is "squeezed" between two opposing grooved fittings or mating parts. Also, O-rings are forced against the walls of the seal by internal pressure making for a very tight seal.

Oxidation   Loss of electrons from a chemical species. Any process which increases the proportion of electronegative constituents in a compound such as increasing the proportion of oxygen in organic compounds.

Passivation   Changing of the chemically active surface of a metal to a much less active state.

Performance Number   Primarily used to rate the antiknock values of aviation gasolines with octane numbers over 100. This value is expressed as the maximum knock-free power output obtained from fuel expressed as a percentage of the power obtainable from isooctane. The relationship between octane number and performance number is listed as follows:

$$\text{Octane Number} = 100 + \frac{\text{Performance Number} - 100}{3}$$

Pilot Pressure   Auxiliary pressure used to actuate or control hydraulic system components.

Pipestill   A furnace containing a series of pipes through which oil is pumped and heated. During heating, the oil is vaporized prior to introduction into the distillation unit or thermal cracking unit.

Polymerization   The process of producing higher molecular weight compounds by reacting two or more unsaturated molecules together. In the refinery, propylene and butylene are the primary polymerization feedstocks.

Poppet   A part of certain valves which prevents flow when the valve closes against a seat.

Positive Displacement   A characteristic of a pump or motor which has the inlet port sealed from the outlet so that fluid cannot recirculate within the system.

Potentiometer   A component of a servo system which measures and controls electrical potential.

Power   The time rate at which work is done.

Pressure   Force applied to, or distributed over, a surface and measured as force per unit area.
   Absolute pressure - pressure measured with respect to zero pressure
   Gauge pressure - pressure measured with respect to that of the atmosphere

Pyrrole   A nitrogen containing, heterocyclic compound found at low concentrations in crude. Pyrrole can readily polymerize in the presence of light to form dark brown compounds. Pyrrole derivatives, such as indole, have been identified as sources which can lead to darkening of fuel color and eventual deposit formation.

Pyrrole

Raffinate   A term used to describe the material recovered from an extraction process. Examples of refinery raffinates include lubestock recovered from furfural or methylpyrrolidone extraction and kerosene recovered from SO_2 extraction. Furfural, methylpyrrolidone and SO_2 extraction processes are employed to remove aromatic compounds.

Range Oil A term sometimes used to describe kerosene or #1 fuel oil.

Rankine Temperature Scale A temperature scale with the size of degree equal to that of the Fahrenheit scale and zero at absolute zero. Therefore $0°R = -459.67°F$ and the normal boiling point of water is $671.67°R$.

Red Dye These dyes are dilutions of C.I. Solvent Red 164 utilized throughout the petroleum industry to identify specific fuels and oils. The dyes are termed *"2-Naphthalenol (phenylazo) azo alkyl derivatives."*

Reduced Crude This term is used to describe the residual material remaining after crude oil has been distilled to remove lighter, more volatile fractions.

Refining Catalyst Most catalysts used in the refining industry are solid heterogeneous catalysts. The chemical reactions they enhance would not proceed at all, or do so quite slowly, in the absence of the catalyst. Reactions of this nature are believed to occur due to a dramatic disruption of the existing chemical bonds of an absorbed molecule. Molecules or molecule fragments may enter into reactions much different from those which occur in uncatalyzed reactions.

Examples of heterogeneous catalysts used in the refining industry are shown in the following table:

Catalyst	Use in Refining
Silica alumina gel	Cracking of heavy petroleum fractions
Chromic oxide gel, chromium on aluminum, nickel-aluminum oxide	Hydrogenation; dehydrogenation of hydrocarbons
Phosphoric acid on kieselguhr	Polymerization of olefins
Co, ThO_2, MgO on kieselguhr (Fischer-Tropsch catalyst)	Synthesis of hydrocarbons from H_2 and CO
Platinum	Isomerization of hydrocarbons
$Al(C_2H_5)_3$; $TiCl_4$	Polymerization of olefins

Note: Kieselguhr is a fine, variously colored earth derived from the accumulated deposits of cell walls of diatoms.

Reforming Both thermal and catalytic processes are utilized to convert naphtha fractions into high octane aromatic compounds. Thermal reforming is utilized to convert heavy naphthas into gasoline quality aromatics. Catalytic reforming is utilized to convert straight-run naphtha fractions into aromatics. Catalysts utilized include oxides of aluminum, chromium, cobalt and molybdenum as well as platinum based catalysts.

Relief Valve A relief valve is used in pumps, reactors and equipment which create positive pressure. These valves are usually connected between the pump outlet and a vent or a large volume accumulation vessel. The valve is normally closed, but will open whenever the system pressure exceeds the desired maximum operating pressure.

Rotary Actuator   A device for converting hydraulic energy into rotary motion.

Ryton   A polyphenylene sulfide elastomer.

Sensitivity   The difference between the research octane number (RON) and the motor octane number (MON) of a gasoline. Fuels blended with alkylate material have low sensitivity values.

Sequence Valve   A sequence valve will permit flow to a secondary system only after an action has been completed in a primary system.

Servo Mechanism   A mechanism which responds to the action of a controlling device and operates as if it were directly actuated by the controlling device. However, it is capable of supplying power output much greater than the original controlling device. The power is derived from an external and independent source. Both mechanical and electrical servo units exist.

Silicone Seal   This elastomer is one of the original materials used as sealing material. It has a wider useful temperature range than Buna-N, but unlike Buna-N, it is not tough and cannot be used in reciprocating systems. Silicone seals tend to swell and absorb oil when operating at high temperatures. Silicone seals are useful in applications where temperatures range from -60°F to 400°F (-51°C to 204°C).

Smoke Point   A measurement of the burning quality jet fuel, kerosene and lamp oils. This value is determined by ASTM D-1322.

Space Velocity   The volume or weight of a gas or liquid which flows through a catalyst bed or reaction zone per unit time divided by the volume or weight of catalyst. The terms liquid hourly space velocity, LHSV, and weight hourly space velocity, WHSV, are used to describe this type of flow measurement. High values correspond to short reaction times.

Specific Gravity   Defined as the ratio of the mass of a given volume of liquid at 15°C (60°F) to the mass of an equal volume of pure water at the same temperature. Typical specific gravity values for some common compounds are listed below.

Compound	Specific Gravity
Water	1.0
Gasoline	0.730 - 0.760
Jet Fuel (Aviation Turbine Kerosene)	0.755 - 0.840
#1 Diesel Fuel and Fuel Oil	0.790 - 0.820
#2 Diesel Fuel and Fuel Oil	0.830 - 0.860
#6 Fuel Oil and Residual Fuels	0.890 - 0.930

Specific Heat Three different values for specific heat can be determined and are described as the *specific heat capacity, specific heat of fusion* and *specific heat of evaporation.* These values are described as follows:

Specific Heat Capacity—Defined as the quantity of heat in Joules required to raise the temperature of 1 kg of a substance by 1°K. For gases, it is necessary to differentiate between specific heat capacity at constant pressure and constant volume.

Specific Heat of Fusion—Defined as the quantity of heat in Joules required to transform 1 kg of a substance at fusion temperature from the solid to the liquid state.

Specific Heat of Evaporation—Defined as the quantity of heat in Joules required to evaporate 1 kg of a liquid at boiling temperature. The specific heat of evaporation is dependent upon pressure.

Petroleum Product	Specific Heat Capacity @ 20°C, kJ./kg•°K
Gasoline	2.02
Diesel Fuel	2.05
Heating Oil	2.07
Kerosene	2.16
Lubricating Oil	2.09
Ethanol	2.43
Water	4.18

Spool A term applied to most any moving cylindrical part of a hydraulic component which directs flow through the system.

Static Dissipater During transport and flow, static charge can build up in fuel and lead to electrical discharge which can ignite fuel/air vapors. This phenomenon can occur in low conductivity fuels such as jet fuel and low sulfur diesel fuel when pumped and filtered at high flow rates. Polar additives known as static dissipaters or electrical conductivity improvers are added to these fuels to promote the transfer of static charge through the fuel to a ground source.

Straight Run Gasoline This stream is the gasoline fraction which is collected from the distillation tower. It is primarily paraffinic in nature and typically has a lower octane number rating than cracked gasoline from the same crude oil feed.

Stripping The separation of the more volatile components of a liquid mixture from the less volatile components. The less volatile components are collected in a relatively pure state, but the more volatile components collected may not be as pure. Stripping may be performed through fractional distillation or by contacting a liquid mixture with a vapor or gas into which the lighter material will diffuse.

Surface Viscosity Molecular films on liquid surfaces may be either readily mobile or slow to flow under the action of a two dimensional stress. Some surface films show surface plasticity and behave as solids until a critical stress is applied. These films exhibit surface viscosity.

Surfactants Soluble compounds which possess groups of opposite polarity and solubilizing tendencies. They can form oriented monolayers at phase interfaces, form micelles, and possess detergency, foaming, wetting, emulsifying and dispersing properties.

Tall Oil Fatty acids and some unsaponifiable material obtained as a major by-product from the sulfite-pulping process of wood.

Theoretical Plate In a distillation column, it is a plate onto which perfect liquid-vapor contact occurs so that the two streams leaving are in equilibrium. It is used to measure and rate the efficiency of a column at separating compounds. The ratio of the number of theoretical plates to the actual number of plates required to perform a separation is used to rate the efficiency of a distillation column. Actual separation trays in refinery distillation units are usually less effective than theoretical plates.

Thermal Expansion Coefficient for Liquids The coefficient of volume expansion for liquids is the ratio of the change in volume per degree to the volume at 0°C. The value of the coefficient varies with temperature.

Three-Way Catalytic Converter A container in-line within the gasoline engine exhaust system which contains platinum, palladium and/or rhodium catalysts. These catalysts convert CO, NO_x and unburned fuel to CO_2, H_2O and N_2.

Topping A process of removing light products from crude oil by distillation. Heavy products remain in the column.

Turbocharging This process utilizes the energy supplied by the combustion exhaust gases to power a turbine which compresses the intake air. The compressed air is more dense and consequently, higher in oxygen content. When delivered to the combustion chamber, compressed air improves fuel economy and engine performance.

Turbine A rotary device that is actuated by the impact of a moving fluid or gas against blades or vanes.

Turbulent Flow A condition where the fluid mass moves in random paths rather than in continuous parallel paths.

Turpentine Light, volatile oils obtained as exudates or from the distillation of rosin oils. It is a mixture of cyclic terpenes composed mostly of α-pinene and has a typical boiling point of about 150°C.

Two-Stroke Cycle Engine An engine which produces one power stroke for each crankshaft revolution. Fuel combustion proceeds as follows:

1. *Air Intake* - With air transfer and exhaust ports open, air under

slight pressure in the crankcase flows into the engine cylinder.

2. *Compression* - The rising piston covers ports, compresses air in the cylinder and creates suction in the crankcase. Fuel injection and combustion is initiated.

3. *Expansion* - Burning fuel expands, pushing the piston down. Air flows into the crankcase to be compressed as the piston descends.

4. *Exhaust* - The descending piston uncovers the exhaust port. A slight pressure builds up in the crankcase enough to move air into the cylinders.

Vapor Pressure   The pressure created in a closed vessel when a solid or liquid is in equilibrium with its own vapor. The vapor pressure is a function of the substance and the temperature.

Volatility Factor   A rating of gasoline quality under environmental and operating conditions. It is a function of the Reid vapor pressure, % distilled at 158°F and the % distilled at 212°F. It can be used to help predict the vapor locking tendency of gasoline.

Water White   A description of the color of petroleum products which is equivalent to rating of +21 or lighter in the Saybolt color scale. Water white fuel components appear as clear liquids to the eye.

Wax Tailings   A pitch-like substance which is the last volatile product distilling from vacuum charge prior to coking.

Weathered Crude   A term applied to crude oil which has lost an appreciable amount of volatile components due to evaporation and other conditions of storage and handling.

Wetting Agent   Surface active compounds such as detergents usually consisting of molecules possessing an oil-attracting hydrophobic group and a water attracting hydrophilic group such as a carboxylic or sulfonic acid group. These compounds act to decrease the contact angle between water and another surface, thereby lowering the surface tension to permit the wetting of the surface.

Wick Char   The weight of deposit which forms on a wick after burning kerosene or lamp oil under specified test conditions. It is a rating of burning quality.

Wiese Formula   An empirical formula developed by General Motors Research for expressing motor fuel antiknock values above 100. This value is obtained by measuring the number of milliliters of tetraethyl lead added per gallon to achieve the same knocking tendency as the fuel being tested. The determined value is then expressed as a *performance number.*

Zeolites Aluminosilicates containing a $(SiAl)_n O_{2n}$ matrix with a negative charge which is balanced by the presence of cations in the cavities. The cations are easily exchanged and water and gases can be selectively absorbed into the cavities. Zeolites are also used to remove molecules of specific sizes by absorption into the pores.

Useful Calculations, Conversions, and Equations

Determination of specific gravity from °API gravity

$$\text{Specific Gravity} = \frac{141.5}{131.5 + \text{API @ 60°F}}$$

Approximation of parts per million (ppm) value from pounds/1000 bbl (ptb) value
- For gasoline 1 ptb = 4.0 ppm
- For kerosene/diesel 1 ptb = 3.3 ppm
- For residual fuel 1 ptb = 3.0 ppm

Conversion Factors
- 1 cu ft = 28.317 liters
- 1 cu ft = 7.4805 gallons
- 1 gallon = 3785.4 milliliters
- 0°F = 459.67° Rankine
- Density of Water @ 60°F = 8.337 lb/gal

1 kg • cal = 1.8 BTU/lb

Calculated Values

Density of Liquid @ 60°F and Atmospheric Pressure

Pounds/gallon = Specific gravity × 8.3372 pounds/gallon

Density of Gas @ 60°F and Atmospheric Pressure

$$\text{Specific gravity} = \frac{\text{Molecular weight}}{28.964}$$

$$\text{Cubic foot vapor/gallon liquid} = \frac{\text{Pounds/gallon} \times 379.49}{\text{Molecular weight}}$$

About the Author

Kim B. Peyton began his career in the petroleum industry in 1978. During the past 19 years, he has worked in the development of synthetic fuels from coal and shale oil, the manufacture and quality assurance of automotive and diesel engine oils, the development of industrial lubricants, the evaluation and development of fuel and lubricant additives and in providing a diverse range of technical support services to customers.

For the past ten years, he has been a Research Group Leader for Nalco/Exxon Energy Chemicals, L.P. located in Sugar Land, Texas. He works daily with laboratory chemists to provide technical service and support to a variety of customers worldwide such as major oil refiners, independent fuel marketers, aftermarket additive suppliers, fuel brokers and others in the petroleum industry.

He has also held technical positions at Union Oil of California (Unocal), Brea, California and at Ashland/Valvoline, Ashland, Kentucky. Additionally, he has earned a B.S. Degree in Chemistry from Marshall University and a M.S. Degree in Organic Chemistry from California State University, Fullerton.

Index

Acid carryover, 68, 179, 203
Acid formation, and fuel sulfur, 112
Acid initiated corrosion, 203
Acid producing bacteria, 101
Active sulfur, 115
Additive induced combustion chamber deposits, 158
Additives, in gasoline, 38–39
Adherent gum, 119, 139
Admiralty brass, 218
Adsorption, in fuel demulsification, 141
Aged gasoline, 89
Air driven pump, 227
Air entrainment, 148
Air sparging, of fuel containing hydrogen sulfide, 115
Alkoxylated compounds, 142
Alkyl aromatics, 16
Alkyl carbamate compounds, 145
Alkyl hydroperoxide, 96
Alkyl lead compounds, 116
Alkyl naphthalene sulfonic acid, 142
Alkyl phenol-formaldyhe compounds, 142
Alkylate:
  in acid carryover problems, 68
  in aviation gasoline, 41
  in fuel corrosion, 177, 203
  in gasoline, 33–34
  influence on induction time, 259
  isobutane as feedstock, 15
  octane number, 37
  production, 17
  vapor pressure, 37
Alkylate gasoline, 34
Alloying elements in steel, 214
Aluminum:
  brass, 216
  cladding, 220
  engine parts containing, 104
  properties, 218
Ammonia, 16, 20, 215
  as a bacterial decomposition product, 101
  oxidation to nitrate by bacteria, 101
Amoco pumpability test for distillate fuel, 186–187
Amsterdam pour point test, 189
Amyl nitrate, 156
Anaerobic species, 142
Analysis of distillate fuel by mass spectroscopy, 54

Anode, 149–152
Antiknock index, 35, 40
Antiknock performance, of gasoline, 33–35
Antiknock properties, of aviation gasoline, 42
Antioxidant:
  compounds and use, 133–135, 138–139
  effect on existent gum, 259
  effect on induction time, 259
  in fuel detergents, 158
  in jet fuel, 46
  problems in use of, 162
  testing, 169–170
Anti-static additive, in jet fuel, 47
°API volume reduction tables, 279–284
Applying corrosion inhibitors, 154–155
Architectural bronze, 216
Arctic grade diesel fuel oil, 51
Aromatic compounds:
  in crude oil, 28
  effect of high concentration, 116–121, 246–247
  effect on burner fuel properties, 59
  effect on cetane number, 189
  effect on gasoline engine knock, 124
  effect on gasoline sensitivity, 37
  effect on specific gravity, 248–249
  formed during FCC processing, 10–11
  formed during reforming, 16
  in fuel stability, 199–200
  in gasoline, 33–34
  in JFTOT, 207
  KB value of, 120
Aromatic content:
  effect on fuel BTU value, 118
  effect on fuel solvency, 120
Aromatic fuel components, swelling of elastomeric seals, 119–120
Aromatic hydrocarbons, cetane engine performance, 117
Ash content:
  of marine fuels, 55
  of residual oils, 62–63
Ash from petroleum products, 191
Asphaltenes:
  description, 29–30
  incompatibility problems, 120
  influence on pour point, 196–197
  influence on sediment, 199–200

Asphaltenes (*Cont.*):
  in marine fuels, 57–58, 120
  in visbreaking, 9
Assessing fuel stability, oxygen overpressure, 174
ASTM D-86:
  to detect contamination of fuel by lubricating oil, 205
  to detect high molecular weight components, 200
  to detect high viscosity components, 197, 200
  to detect low viscosity components, 198
  to help resolve combustion quality problems, 257
  to identify gasoline T-90 limits, 130
  to predict deposit forming tendency, 199
  to predict diesel low temperature problems, 185
  to predict fuel sediment problems, 200, 257
  to predict kerosene flame height problems, 200
ASTM D-97, 182–183, 187–189, 193
ASTM D-130, 175
ASTM D-381, 170–171
ASTM D-396, 51
ASTM D-482, 191
ASTM D-525, 169–170
ASTM D-613, 93, 189
ASTM D-665, 177
ASTM D-873, 191
ASTM D-910, 43
ASTM D-975, 50
ASTM D-1094, 43, 178
ASTM D-1319, 190, 199, 207
ASTM D-1401, 179–180
ASTM D-1500, 171
ASTM D-2274, 138, 173
ASTM D-2276, 68, 208
ASTM D-2500, 183
ASTM D-2699, 33, 35
ASTM D-2700, 33, 35, 42
ASTM D-3120, 190
ASTM D-3227, 190
ASTM D-3703, 191
ASTM D-3948, 180–181
ASTM D-4176, 181–182
ASTM D-4539, 81, 185
ASTM D-4625, 172–173
ASTM D-4814, 169–171, 175
ASTM D-5304, 138, 174
ASTM D-5500, 41
ASTM D-5598, 41
Atmospheric distillation, 6
Autoignition:
  in cetane engine, 189
  of diesel fuel, 90–94
  early ignition, 125
  effect of cetane improver, 93–94
  important factors, 252
Auxochromic groups, 98
Aviation fuel additives, 46–47
Aviation fuel grades, 44
Aviation gasoline, properties, 42–43

Aviation turbine fuel (*see* Jet fuel)

Ball-on-cylinder lubricity evaluator, 114, 121, 160–161
Basic forms of corrosion, 152–153
β-oxidation, 142
β-scission, 12
Biobor JF, 47
Biological oxygen demand (BOD), 66
Black smoke, 121
Blue smoke, 103
BMW, 41, 159
BOCLE (*see* Ball-on-cylinder lubricity evaluator)
Boron, as a fuel contaminant, 105
Boundary lubrication, 160
Brass, types of, 216
Break point, of gasoline, 169–170
British admiralty pour point test, 188
Bronze, types of, 217
BTU ratings per pound, 273
Bunker fuel, 51, 56, 61
Burner fuel, properties, 58–59
Butylene oxide, 142

C factor, 155
Calcium:
  chloride, 22
  in crude oil, 33
  in engine oil, 103–104
  salt formation with corrosion inhibitor, 68
  salts in unwashed and washed gums, 171
  in sea water, 105
  in tank bottom water, 139
Calculated carbon aromaticity index (CCAI), of marine fuels, 55, 57
Calculated ignition index (CII), of marine fuels, 55, 57
Carbazoles, effect on fuel color, 99
Carbon-carbon reactions, formation of soot, 124
Carbon deposits, 10, 126–127
Carbon dioxide, from fuel refining, 20
Carbon residue, 53, 251
Carbon steel, 211
Carburetor and fuel injector detergents, 158
Carburetor icing, 76
Catalyst fines:
  description, 102
  effect on fuel quality and performance, 102
Catalyst poisons, 14, 16
Catalysts:
  alkylation, 17, 68, 201
  combustion, 157
  deactivation by metals, 26, 32
  fines in marine fuel, 55, 62
  in fuel degradation, 137, 173
  hydrocracking, 14
  hydrotreating, 13

Catalysts (*Cont.*):
  reforming, 16
  sweetening, 21
Catalytic hydrocracking, 13–14
Catalytic polymerization, 15
Catalytic reforming, 15–16
Caustic:
  in acid removal, 68, 250
  carryover, 68, 179, 202, 207–208
  effect on acidic corrosion inhibitor, 223–225
  effect on copper, 215
  in emulsification, 140
  in fuel sweetening, 156
  processing of light straight run gasoline, 34
  role in filter plugging, 163, 208
  to improve induction time, 260
  to remove existent gum, 259
  to remove sulfuric acid, 208
  treatment of FCC propane-butane, 10
  use in ASTM D-1094 testing, 179
  use in naphthenic acid removal, 207
  washing, as a refining process, 19, 21
Centrifuge sludge, due to asphaltene incompatibility, 120
n-Cetane (*see* Cetane number)
Cetane engine reference fuels, 95
Cetane improver:
  compatibility with elastomers, 166–167
  concentration in federal diesel fuel oil, 92
  description, 156–157
  effect on carbon number, 127
  effect on stability test results, 166
  realistic performance, 91–92
Cetane index, 52, 90–92, 119
Cetane number, 52
  determination, 90–95
  high sulfur/low sulfur difference, 49
  low, 261
  reproducibility, 93
CFPP (*see* Cold filter plugging point)
Chemical oxygen demand, 66
Chemical storage tanks, 222–223
Chemistry of demulsifiers and dehazers, 142
Chromium:
  in crude oil, 33
  engine parts containing, 104
  in steel, 213–214
Chromophoric groups, 98
Cladosporium resinae, 100
Clarified oil, 10
Clay filtration:
  of jet fuel, 46, 206
  to remove color bodies, 199, 261
  to remove copper, 207
  to remove existent gum, 259
Clostridia, 101–142
Cloud point improver, 81, 165, 205
Cloud point of petroleum products, test method, 183

Cloud point reduction, 81–82
Cloudy or hazy, due to low temperature, 74
Coalescence:
  electrostatic, 23
  in fuel demulsification, 141
Cobalt-molybdenum (CoMo), 13
COD (*see* Chemical oxygen demand)
Coefficient of expansion, 70
Coker gas oil:
  effect on carbon number, 126, 251
  production, 18
  role in fuel stability, 200
  specific gravity influence, 249
Coking, in refining, 18–19
Cold filter plugging point, 83–84, 87, 183–184
  antagonism by cloud point improver, 165–166
  as a filter plugging test method, 81
  to identify wax related problems, 193–194
  reduction by kerosene, 83–84
  test method, 183–184
Cold flow improvers (*see* Wax crystal modifiers)
Cold weather engine starting, 128
Cold weather engine warm-up, 128
Colloidal ferric hydroxide, 67, 150
Colonial pipeline haze rating, 181
Color bodies, 20, 22, 77, 98–99, 113, 199–200
Color change, effect of distillate fuel stabilizer, 139
Combustion chamber:
  deposits:
    effect of alkyl antiknock compounds on, 116
    effect of fuel sulfur on, 53, 114
    effect of low speed on, 114
    effect on octane number requirement, 36
    in gasoline engines, 41
    influence on cylinder temperature and pressure, 125
  effect of cold temperature on, 76
  fuel vaporization in, 88–90
  injection of fuel into, 232
  oil leakage into, 103
  poor mixing of fuel in, 252–253
  pressure, 125
    on piston ring, 129
  sources of, 157–159, 257
Components used to prepare gasoline blends, 34
Condensed aromatics, and combustion chamber deposits, 41
Condensed water, 75–76
Contaminants from refining processes, 20
Copper, engine parts containing, 104
Copper alloys, 215–217
Copper chloride sweetening, 21
Copper corrosion, 13, 156, 175, 208–209, 260
Copper ions, 137, 139, 200
Copper mercaptide, 260
Copper strip tarnish test, 174–175
Copper sulfide, 112, 155, 208
Corrosion chemistry, 149–151

Corrosion inhibiting properties of fuel stabilizers, 136–137
Corrosion inhibitor, 148–155
  in jet fuel, 47
  salts, 69, 208–209
  use in system containing rust, 154
Corrosion of aluminum, 219–220
Corrosion rate of steel, 150–151
Cracked gasoline, 34
Cracking:
  by catalytic hydrocracking, 13–14
  by coking, 18–19
  by fluid catalytic cracking, 9–11
  fuel contaminants from, 20
  of metal under hotspots, 53–251
  prevention by vacuum distillation, 7
  stress cracking corrosion of steel, 153, 214
  by thermal cracking, 8–9
  by visbreaking, 9
Cresols, effect of exposure to light, 78
Crevice corrosion, 152
Crude oil:
  asphaltenes, 29–30
  assay, 25–26
  composition, 27–33
  compounds affecting fuel color, 199–200
  hydrocarbons, 27–28
  nitrogen in, 32
  metals in, 32–33
  oxygen in, 32
  pour point reversion, 87–88, 196–197
  pour point testing, 187–189
  refining, 5–19
  residual fuel components, 60–61
  shipping information, 236
  sulfur, 31, 246
  wax, 79–80
Crystallization, of antioxidants, 162
Cummins L-10, 160
Cupronickel alloys, 216
Cyclohexyl nitrate, 156
Cyclone, 10
Cycloparaffins, 16, 27

Daimler-Benz 2.3L M 102E, 159
Darkening of refined fuels, 199–201
Debutanizer, 17
Dehazing, 140–142
Dehydrocyclization, 16
Dehydrogenation, 8, 16
Delayed coking, 18
Demulsification, 140–142
Demulsifiers and dehazers, 140–142
Deposit analysis, 253–254
Deposits in combustion chamber, 257
Deposits in fuel systems, 157–159

Desalting, 5
Desulfomonas, 100, 142
Desulfotomaculum, 100
Desulfovibrio, 100, 142
Desulfurization, 6–7, 13
Detergent:
  additive use, 39, 157–160, 256–257
  interaction with demulsifiers, 69, 140–142, 164
  testing in engines, 41
Diaphragm pump, 227
Diaromatic and polyaromatic compounds, lubricity effect, 121
Dibrominated proprionamide, 144
Diesel fuel:
  additive related problems, 163–167
  appearance testing, 181–182
  aromatics, 116–120
  carbon number, 126–127
  cetane:
    engine testing, 90–95, 189–190
    improver, 91–92, 156–157, 166–167
    index, 90–91, 119
    number, 90–95, 117
  cloud point, 81–82, 183
  cold filter plugging point, 83–84, 183–184
  cold temperature problems, 73–76
  color degradation, 199–201
  combustion, 251–252, 255
  components
    from catalytic hydrocracking, 13–14
    from distillation, 7
    from fluid catalytic cracking, 10
  copper corrosion testing, 175
  density:
    and effect on BTU rating, 198
    and fuel consumption, 118
  deposits, 73, 154–157
  detergents, 157–160
  emulsification problems, 68–69, 141
  filter plugging point testing, 183–187
  filters, 229–230
  grades, 49–51
  hot restart problems, 111–112
  ignition delay, 93–94
  injection pump sticking, 163
  injection pumps, 230–232
  low temperature flow test, 185
  LTFT reduction, 84
  lubricity, 113–114, 121
  lubricity improver and testing, 160–161
  marine, 54–58
  naphthenes, 27
  paraffins and cetane number, 122–123
  pour point testing, 182–183
  power loss, 197–198
  pour point reduction, 82–83
  properties, 52–54

Diesel fuel (*Cont.*):
  pumpability testing, 186–187
  pumping problems, 193–195
  sediment formation, 256–257
  smoke, 103, 121, 124
  stabilizer:
    composition, 136–139
    federal, 138
    naval distillate, 138
  startability, 129
  sulfur combustion, 112–115
  sulfur related problems, 246
  tests to determine filter plugging tendency, 81
  viscosity related problems, 107–112, 245, 249–250
  volumetric fuel consumption, 118
  warmup, 128
  water reaction testing, 178–179
  wax, 80–81, 123
  wax crystal modifier problems, 85–87, 165–166
  wax crystal modifiers, 145–148
Diester-alphaolefin copolymers, 145
Diethyleneglycol monomethylether, (DEGME), 181
Dimersol process, 15
Diolefins, 20, 158, 206
N,N'-Disalicylidene-1,2-cyclohexanediamine, 137
N,N'-Disalicylidene-1,2-propanediamine, 137–138
Dispersant, property of distillate fuel stabilizer, 136
Disperse wax, 194
Distillate fuel degradation, 136–139
Distillate fuel stability, testing for, 171–174
Distillate fuel stabilizers, 136–139, 166, 199–200
Distillation:
  atmospheric, 5–6
  fractional, 25
  fuel properties determined by, 128–131
  profile to predict performance, 52, 191, 251
  of residual fuel, 63
  vacuum, 6–7
Distributor pump or rotary fuel pump, 230–231
Disulfides, 115
Disulfovibrio, 100
DuPont F21 stability test, 172
Dyes, 39

EGME (*see* Ethyleneglycol monomethylether)
Elastomer compatibility, with cetane improver, 166–167
Elastomer degradation, 113
Elastomers, 40, 119, 221–222
Electrical conductivity, 46, 49, 249
Electrical service, 229
Elemental sulfur, 115, 155, 174, 208
Emissions, 41, 53, 57, 94–95, 128–129
Emulsification:
  of additives in jet fuel, testing for, 206
  due to cloud point improver, 205
  due to naphthenic acid salts, 140, 201

Emulsification (*Cont.*):
  of polar organic compounds, 204
  problem sources, 139–140
  testing, 177–182
Emulsions:
  created by caustic carryover, 68
  in crude oil, 5
  due to lubricating oil, 205
  due to microorganisms, 204–205
  due to polar organic compounds, 204
  prevention by demulsifiers, 141–142
  resolving and preventing, 258–259
  stabilized by organic acids, 250
Engine efficiency, effect of intake air temperature, 70–71
Engine oil contamination, of fuel, 103–104
Engine tests, 41
Enjay fluidity, 81, 185–186
EPA registration, of biocides in U.S., 143
Epoxide, 142
Ethylene oxide, 142
Ethylene vinylacetate, 85, 145–146
Ethyleneglycol monomethylether, 45, 47, 81, 181
2-Ethylhexyl nitrate, 156
  physical properties, 157
EVA (*see* Ethylene vinylacetate)
Exhaust emissions, 41
Exhaust odor, 112–113
Exhaust valve seat wear, 116
Existent gum, 38, 40, 170–171, 259
Existent gum in fuels by jet evaporation, 170–171

Fatty acid type corrosion inhibitor, structure, 154
Fatty acids, from fuel refining, 20
Fatty amines, 160
FCC (*see* Fluid catalytic cracking)
Federal diesel fuel oil, 138
Ferric hydroxide, 67, 101, 150
Ferric oxide, 150
Ferrous hydroxide, 150
Ferrous ions, 149–150
Ferrous metal corrosion, 67–68, 149–150, 176–177, 202–204, 258
FeS, 101, 115, 152
15:5 Distillation, 25
Film forming, 153–154, 163
Filming inhibitors, 153–154
Filter plugging:
  by corrosion inhibitor salts, 154, 163
  by detergent emulsions, 140–141
  by fuel wax, 84, 145, 183–186, 193
  by jet fuel particulates, 207–208
  by microorganisms, 101
  by organic salts and emulsion, 68–69
  by pipeline rouge (rust), 163
  by sediment in marine fuel, 57–58

Filters, diesel fuel, 229–230
Filtration, in fuel refining, 23
Finishing processes, 19–24
Flexicoking, 19
Flowmeters, 227–228
Fluid catalytic cracking, 9–12
Fluid coking, 18–19
Fluorenes, effect on fuel color, 99
Forging brass, 216
Free radical:
  inhibition by antioxidants, 133
  light activated reactions, 77–78
  oxidation, 96–97, 113
  in thermal cracking reactions, 8
Fuel efficiency:
  effect of cetane number, 94–95
  effect of viscosity, 58, 107–110, 245–246
Fuel filterability, 183–185
Fuel gas, 7
Fuel induced deposits, 158
Fuel injection timing, possible effect on smoke and hydrocarbon emissions, 95
Fuel odor, 114, 190
Fuel oil:
  biocides, 142–144
  burner applications, 58–60
  federal stabilizer for, 138
  grades, 49–51
  heating value loss, 128
  properties, 52–54
  marine, 54–58
  residual, 60–64
  viscosity limits, 108
Fuel storage and transportation system corrosion, 67
Fuel sulfur:
  effect on catalytic converter, 116
  effect on combustion chamber deposits, 114
  effect on combustion system corrosion, 112
  effect on diesel fuel lubricity, 113–114
  effect on exhaust odor, 112–113
  effect on lead antiknock compounds, 116
Fuel sweetening additives, 156
Fuel system icing inhibitor, in jet fuel, 47
Fuel viscosity:
  effect of cold temperature on, 75
  effect of high viscosity, 107–110
  effect of low viscosity, 110–112
  effect on atomization, 109
  effect on burner fuel performance, 59
  effect on filling of fuel pump, 107
  effect on fuel combustion, 197–198
  effect on kerosene flame height, 109
  effects, 245–246, 249, 250
  problem solutions and preventions, 255–256
  relationship to diesel fuel IBP and EP, 194
Fuel volatility, effect of low temperature, 76–77
Fugitive emissions, 41

Gallionella, 101
Galvanic corrosion, 152
Galvanic series of metals and alloys, 285
Gas oil:
  coker, 18
  as FCC feedstock, 9
  marine, 56
  in residual fuel, 60, 187
  vacuum, 7
Gasohol, 40
Gasoline:
  additives, 38–39
  alkyl lead antiknock compounds, 116
  antiknock performance, 33, 35–36
  antioxidants, 133–135
  API gravity range, 119
  aviation, 41–43
  biocide registration, 143
  blending components, 34
  carburetor deposits, 157
  catalytic converter efficiency, 116
  combustion chamber deposits, 41, 129, 257
  combustion summary, 252–253
  components:
    from alkylation, 17
    from atmospheric distillation, 6
    from catalytic hydrocracking, 13–14
    from catalytic polymerization, 15
    from catalytic reforming, 15–16
    from fluid catalytic cracking, 10
    from isomerization, 15
    from thermal cracking, 8
    from visbreaking, 9
  contamination, 126
  copper corrosion testing, 174–175
  crankcase dilution, 129
  demulsibility testing, 178–180
  demulsification, 139–142
  detergent additive emulsification, 69
  detergents, 157–159
  effect of:
    acid carryover, 68, 201
    aging, 88–89
    caustic washing on, 19, 21
    ethanol blending, 40
    low temperature on, 76
    olefins, 97
    percent distilled at 158°F on engine warmup, 89–90
  effect on distillate fuel wax, 194
  engine tests, 41
  exhaust emissions, 41
  existent gum testing, 170–171
  FCC, 34
  gum and sediment, 256–257
  hydrocarbon type testing, 190
  induction time failure, 259–260
  intake valve deposits, 158–159

Gasoline (*Cont.*):
  knock, 124–125, 130–131
  octane number requirement, 36
  oxidation stability testing, 169–170
  oxygenates, 39–40
  power loss, 255–256
  properties, 33–38
  sweetening, 21–22
  T-90 limits, 129–130
  vapor locking, 126
  vapor pressure, 88–89, 126
  volatility, 37, 88, 126
Gear pump, 225
Glutaraldehyde, 144
Good housekeeping practices, 23
Green coke, 18
Gum formation:
  from cracking processes, 19
  from FCC gasoline, 34
  in fuel systems, 157–158
  in gasoline or diesel fuel, 256–257
  mechanism, 96–97
  by oxygenated organics in jet fuel, 45
  prediction of, in gasoline, 169
  prevention by antioxidants, 39, 133–134
  prevention in jet fuel, 46

Hard starting, 253
Hastelloy, 221
Hazard information for common fuels, 233–244
Heat:
  effect on corrosion rate, 71–72
  effect on fuel volume, 69–70
  effect on reaction rate, 71
Heat exchanger, 216, 220
Heat of combustion, 45, 52, 198
Heat tracing, 228
Heating oil, 49–54
Heavy straight run (HSR) gasoline, 6–7, 33
Heptamethylnonane, cetane number, 95
Heptane, as a reference fuel, 35
Heterocycles:
  in crude oil, 30–32
  in distillate fuel, 113
  in fuel darkening, 199–200
Hexahydro-1,3,5-triethyl-*S*-triazine, 144
Hexyl nitrate, 156
HFRR (*see* High frequency reciprocating rig)
High acid number, 250
High aromatic content, 246–247
High BTU, 59, 118
High carbon values, 251
High Conradson carbon number, 126–127, 166, 251
High electrical conductivity, 249
High existent gum, sediment, 256, 259
High frequency reciprocating rig, 160–161

High molecular weight aromatics, 158
High octane gasoline, 13, 15–16
High paraffin content, 247
High pressure atomizing gun burner, 60
High specific gravity, 248–249
High speed engines, 55
High sulfur content, 246
High sulfur diesel, 48–49, 113–114, 246
High viscosity, 245–246
Hindered phenol, 133–134
Homolytic cleavage, 86
Hot restart problems in diesel engines, 111–112
HUM (*see* Hydrocarbon utilizing microbes)
Hydrocarbon types, ASTM test method, 190
Hydrocarbon utilizing microbes, 100
Hydrocarbons in crude oil, 27–28
Hydrocracked gasoline, 13, 34
Hydrocracking, 13–14, 16, 79
Hydrodesulfurization, 12–13, 21, 63
Hydrofluoric acid, 17, 201, 203
Hydrogen abstraction, 96
Hydrogen gas, 13–14, 150, 155
Hydrogen ions, 150
Hydrogen sulfide (H_2S):
  in catalytic reforming, 16
  corrosion of copper, brass, bronze, 112, 155, 208
  corrosion of iron, 151
  in crude oil, 26
  formation during hydroprocessing, 13
  glutaraldehyde degradation by, 144
  in hydrocracking, 14
  product of flexicoking, 19
  product of microbial metabolism, 100, 143
  production in fluid catalytic cracking, 10
  reaction with an olefin, 100
  from refining, 12–13, 16, 19–20
  removal by additives, 156
  removal by molecular sieves, 24
  removal by sweetening, 21–22
  removal from diesel fuel, 48
Hydrogen transfer, 8, 12
Hydroperoxide decomposition, 96, 134, 137
Hydroperoxides, 96–97, 134
Hydroprocessing, 12–13, 48, 260
Hydrotreating, 12–13, 79
Hypalon, 222

Ice, 194
Ignition delay, 93–94
Ignition quality, test method, 189–190
Imadazolines, 30, 154
Indane, 28
Indole, 30, 32, 99, 199
Induction pPeriod, 169–170
Injection pump sticking, 163
Inline injection pumps, 231–232

Intake valve deposit, 41, 158–159
  detergent, 159
  formation, 158
Invar, 220
IP 15, 182
IP 40, 169
IP 41, 189–190
IP 79, 179
IP 94, 191
IP 95, 190
IP 123, 191
IP 131, 170
IP 135, 177
IP 138, 191
IP 154, 175
IP 219, 183
IP 289, 178
IP 309, 81, 183–184
IP 342, 190
IP 378, 172
IP 388, 173
Iron, engine parts containing, 104
Iron sulfide, as a product of microbial metabolism, 101
Isomerate, 33
Isomerization, 15–16
Isooctane, reference fuel, 35
Isooctylnitrate, 91, 156
Iso-paraffins, 9, 15–17, 33
Isopropyl nitrate, 156
Isothiazoline, 144
IVD (*see* Intake valve deposit)

Jet A, 44, 47
Jet A-1, 44, 47
Jet fuel:
  additives, 46–47
  antioxidants, 133–134
  from catalytic hydrocracking, 13–14
  color degradation by PDA, 162
  as a diesel fuel blendstock, 48, 81
  grades, 44
  JFTOT, 46, 135, 206–207
  lubricity, 160
  molecular sieve processing, 24
  particulate contamination, 68, 207–208
  properties, 44–46
  water reaction test, 178–179
  water separation index, modified, 163, 180–181
JFTOT (Jet fuel thermal oxidation test) [*see* Jet fuel]
JP-8, 44

Kati-condensed asphaltenes, 30
Kauri-butanol value (KB value), 120
KB value of selected petroleum products, 122

Kepner-Tregoe problem analysis technique, 2
Kerosene:
  in CFPP reduction, 83–84
  in cloud point reduction, 81–82
  as a diesel fuel blendstock, 48, 81–84, 93
  effect on cetane number, 93
  effect on diesel lubricity, 111
  flame height reduction, 109
  illuminating and burning quality, 90
  Jet A-1 type, 44
  KB value, 122
  in LTFT reduction, 84
  mercaptan sulfur test method, 190
  in pour point reduction, 82–83
  viscosity limits, 108
Knock:
  in aviation gasoline, 42
  due to deposits, 44, 114
  in diesel engine, 94
  in gasoline engine, 124–125, 130–131
  octane number requirement, 36
  prevention by lead compounds, 116
  test methods, 33, 35–36

LCO (*see* Light cycle oil)
Lead, in engine parts, 104
Lead scavengers, 116
Leptothrix species, 101
Light, effect on fuel color, 77–78
Light cycle oil, 10, 136, 199
Light straight run (LSR) gasoline, 6–7, 34
Low aromatic low sulfur diesel, 113–114
Low BTU, 128
Low calorific value, 198
Low flash point, 247–248
Low fuel viscosity, effect on lubricity, 111
Low induction time in gasoline, 259–260
Low octane number, 130–131
Low power, 197–198
Low pressure air atomizing burner, 60
Low specific gravity fuel, 198
Low sulfur diesel, 13, 48–50, 108, 113–114, 127
Low temperature flow test:
  as a filter plugging test, 81
  reduction, 84
  test method, 185
Low viscosity, 249–250
LTFT (*see* Low temperature flow test)
Lubricating oil:
  and combustion chamber deposits, 41, 158
  dilution by fuel, 76–77
Lubricity, 49, 111, 113–114, 121, 160
Lubricity additive, in jet fuel, 47
Lubricity improver, 160–161
Lupolen, 222

Index

Macrocrystalline wax, 80
Magnesium, 33, 40, 62, 105, 139
Maintaining demulsifier solubility within the package, 164
Mal-shaped wax, 79–80
Maltenes, in crude oil, 29
Manganese bronze, 216
Marine diesel oil (MDO), 56
Marine fuel, properties, 54–58
Marine fuel pumping problems, 110
Marine gas oil (MGO), 56
Marine intermediate fuels, 56
Measuring rapid degradation by oxidation, 173–174
Medium speed engines, 55
Mercaptan sulfur, test method, 190
Mercaptans:
  aluminum corrosion resistance to, 218
  copper mercaptide formation, 260
  in crude oil, 31
  in FCC products, 10
  as a fuel contaminant, 20
  in hydroprocessing, 13
  odor, 31
  oxidation, 114–115
  reaction with olefin, 100
  removal, 19, 21–22, 156
Merox process, 21–22
Metal chelation, as a fuel stabilizer property, 136–138
Metal deactivator, 39, 46, 137–138, 158, 170, 259
Metal depositing bacteria, 101
Metal loss, 152–153
Metals, in crude oil, 26, 32–33
Methyl tertiary butyl ether, 34, 40–41, 178
Methylene-bis-thiocyanate, 144
Microbial growth, 100–102, 142
Microbial plaques, 100
Microbiocides, 47, 142–144
Microbiologically influenced corrosion (MIC), 100, 143
Microcrystalline wax, 80
Microorganisms, 66, 100–102, 139, 202, 204
Microseparometer, 180–181
Military jet fuel, 44–47
MIL-S-53021, 138
Mixed polycyclic hydrocarbons, in crude oil, 28
Molecular sieves, 14, 23–24, 206
MON (*see* Motor octane number)
Monel, 220
Monoethanolamine, 10
Motor octane number, 17, 33–37, 42, 130
MSEP (*see* Microseparometer)
MTBE (*see* Methyl tertiary butyl ether)
Muntz metal, 216

NACE corrosion, 176
NACE spindle rating scale, 176

Naphtha:
  in aviation gasoline, 42
  from coking, 18
  from distillation, 6–7
  from hydrocracking, 13
  hydroprocessing, 13
  in jet fuel, 44
  nitrogen in, 32
  oxygen in, 32
  reforming, 15
  sulfur in, 31
Naphthenic acids:
  description, 32
  in distillate fuel, 136
  as emulsifying agents, 140
  from fuel refining, 19
  in jet fuel, 45–46
  reaction with caustic, 202, 207
  removal, 19, 21
Naphthenic-aromatic compounds, in crude oil, 28
Naval bronze, 216
Naval distillate fuel, 56
Naval distillate stabilizer, 138
Needle wax, 80
Nickel:
  alloys, 220–221
  in brass, 216
  in bronze, 217
  in crude oil, 25–26, 32
  as a fuel refining catalyst, 13–14
  in marine fuel, 62
  in steel, 153, 215
Nickel-molybdenum (NiMo), 13
Nitrobacter, 101
Nitrosomonas, 101
Non-positive displacement pumps, 223–224
Normal butane, in gasoline, 33
Nucleation, 81, 145, 165

Octane number:
  of automotive gasoline, 33–37
  of aviation gasoline, 42
  of FCC products, 10–11
  refining processes affecting, 13–17
  requirement (ONR), 36
  values of blendstocks, 37
Octane requirement index, 36, 159
Octyl nitrate, 156
Odor, 45, 53, 57, 112–113, 128
  see also Mercaptans
Oil and grease, in water analysis, 67
Olefins:
  additives to treat effects of, 133–134, 157–160
  as alkylation feedstock, 17
  as catalytic polymerization feedstock, 15
  as chromophores, 98

Olefins (*Cont.*):
  cetane number, 123
  in crude oil, 27
  in deposit formation, 41, 158
  in diesel fuel, 48, 97, 256
  effect on fuel quality, 97
  effect on sludge formation, 97
  effect on varnish formation, 97
  as FCC unit products, 10–11
  in gasoline, 34, 38, 256
  in jet fuel, 44, 206
  reaction with hydrogen sulfide, 100
  removal by acid treatment, 22
  removal by hydroprocessing, 12
ONR (*see* Octane number)
Opel Kadett 1.2L 12S, 159
Operability additives, 84
Organonitrogen compounds, in crude oil, 32
Organosulfur compounds:
  as antioxidants, 134
  in crude oil, 31
ORI (*see* Octane requirement increase)
Oxidation:
  copper catalyzed, 137
  of olefins, 97
Oxidation stability:
  of aviation fuels, 174
  of distillate fuel oil, 173
  of gasoline, 169–170
Oxidative/thermal stability, of low sulfur diesel, 49
Oxygen, and deposit formation, 72–73
Oxygen containing compounds, in crude oil, 32
Oxygenate, 34, 40, 178
Oxygenated organics, in jet fuel, 45

Packing group, 234
Paint, 40, 162
Paraffinic hydrocarbons, effect on cetane number, 123
Paraffins:
  in crude oil, 27
  hydrogen content and fuel combustion, 124
  as solvents, 122
  wax related problems, 123
n-Paraffins, 9, 122, 195
Passive films, 152
PDA (*see* Phenylenediamine)
Percent gasoline distilled at 158°F engine warmup, 89–90
Perforated sleeve vaporizing burner, 60
Performance number, of aviation gasoline, 42
Peri-condensed asphaltenes, 29
Permalloy, 220
Peroxide number of aviation turbine fuel, 191
Peroxy radical, 96
PFI (*see* Port fuel injection)
Phenolic antioxidants, 133–134, 162

Phenols:
  effect of exposure to light, 78
  effect on fuel color, 99
Phenylenediamine, 21, 134–135, 139
Phosphorus content, of gasoline, 39
Piston pump, 226
Pitting, 153
Plaques, 100, 143, 202
Plate wax, 79
Platinum, 14–16
Polycyclic aromatic compounds, 116, 200
Polynuclear aromatics, 9, 28
Polytetrafluoroethylene, 221
Poor flame quality, kerosene, 198–199
Poor startability, 76, 251–253
Poor starting, warm-up and initial acceleration, 157–160
Porphyrin ring, 32
Port-fuel injection, 41
Positive displacement pumps, 224–227
Potential heat, effect of aromatic content, 118
Pour point:
  of petroleum products, 182–183
  reduction of diesel fuel, 82–83
  of residual oils, 62
  reversion, 146, 188, 196–197
Predicting storage stability of distillate fuel, 172–173
Preventing detergent-water emulsions, 141
Products of microbial metabolism, 143
Products produced from catalytic cracking, 10
Propane-butane, 10
Propylene oxide, 142
Pseudomonas species, 101
PTFE (*see* Polytetrafluoroethylene)
Puegeot XUD-9, 160
Pumpability problems, 146, 193–195
Pumps, 223–227, 230–232
Pyridine, 32
Pyrrole, 32

QPL (Qualified Product List)-24682, 138
Quaternary ammonium salts, 142
Quinoline, 32, 78, 199

Reactions which can liberate sulfur, 115
Red brass, 216
Reduced power, 255
Reducing cloud point, 181–182
Reducing pour point, 82–83, 85–88, 165, 182–183
Reformate, 15–16, 33
Reforming (*see* Catalytic reforming)
Reformulated gasolines, 41
Regular grade diesel fuel oil, 51–52
Reid vapor pressure, 37, 40, 88–89, 126
Removal of inhibitor film, 203

Index

Research octane number, 35
Residual fuel:
  asphaltene incompatibility, 120
  coking, 18–19
  as marine fuel, 56–57
  pour point, 88, 145, 187–189
  pour point increase, 195–197
  properties, 60–64
  in visbreaking, 9
  wax, 79–80
Resins, in crude oil, 29
Reversion of pour point, 165, 187–189, 195–197
Riser, 9–10
RON (*see* Research octane number)
Rotor or lobe pump, 225–226
Rouge removal, by corrosion inhibitors, 163
Rust:
  formation process, 150
  in fuel, 78–79
  on metal, 201–202
  prevention, 153–154
  reaction with hydrogen sulfide, 115
  removal by corrosion inhibitor, 153–154, 163
  test methods to predict, 176–177
Ryton, 221

Salt drying, in fuel refining, 22
Saybolt color degradation, due to PDA antioxidants, 162
Sea water contamination, 105
Seal swelling, 119–120
Sediment, determination in marine fuels, 57–58
Sensitivity, in gasoline octane number determination, 36–37
SFPP (*see* Simulated filter plugging point)
Shell-Amsterdam pour point test, 189
Silver, engine parts containing, 104
Simulated filter plugging point, 81, 184–185
Slime forming bacteria, 101
Slow speed engines, 55
Sludge:
  definition, 103
  formation, 96–97
  in fuel, 119–120, 199–200
  inhibition , 136–137
  prevention, 133–139
  tests to predict formation of, 171–174
Slurry oil, 10
Smoke:
  black, 121
  blue, 103
  causes and prevention, 257–258
  from diesel fuel combustion, 95, 128
  formation by burner fuels, 59
  reduction, 252, 257–258
  white, 76

Smoke point, 14, 45, 247
Sodium dimethyl dithiocarbamate/disodium ethylene-bis-dithiocarbamate, 144
Solving a problem, 1–3
Southwest Research scuffing ball-on-cylinder lubricity evaluator, 160
Specific gravity, of fuel, 119
SRB (*see* Sulfate reducing bacteria)
Stabilizer packages, composition, 136
Stainless steel, 214–215
Startability, 52, 129
Steam:
  in coking, 18
  in distillation, 7
  in fluid catalytic cracking, 9–10
  tracing of lines, 228
Steel, 211–214
Straight run naphtha, 7, 15
Stress corrosion cracking, 153
Styrene-ester copolymers, 145
Succinimides, 69, 160
Sulfate reducing bacteria, 100–101, 142
Sulfides, in crude oil, 31
Sulfonates, 112, 154
Sulfonic acids, 115, 140
Sulfur:
  acids, 63, 114
  analysis, 190
  in aviation gasoline, 43
  in copper corrosion, 155, 208
  in crude oil, 26, 31
  in diesel fuel, 58–50, 53
  effect in fuel, 112–116
  effect on copper corrosion, 112
  effect on fuel lubricity, 113–114
  effect on fuel stability, 113
  formation reactions, 115
  in gasoline, 38
  in jet fuel, 45
  in lubricating oil, 104
  in marine fuel, 57
  in residual fuel, 63
  reaction with steel, 214
  reduction by hydroprocessing, 13
Sulfur and nitrogen, in asphaltenes, 29
Sulfur and SO_x, 63
Sulfuric acid:
  as an alkylation catalyst, 17
  carryover, 68
  initiated corrosion, 201–203, 209
  role in deposit formation, 113–114
  stainless steel resistance to, 214
  from sulfur trioxide, 63
  from *Thiobacillus*, 101
  treating of fuel, 22
Sweetening processes, in fuel refining, 21–22
Synergistic effect, of antioxidants, 133

T Fuel, cetane number, 95
T Fuel and U Fuel, low temperature storage, 95
Tensile stress, 153
Terne, 229
Tertiary-alkyl primary amines, 138
Tetraethyl lead, 43, 47, 116
Tetralin, 28
Thermal cracking, 8–9, 20
Thiobacillus thiooxidans, 101
Thiols or mercaptans, in crude oil, 31
Thiophenes, in crude oil, 31
Tin, engine parts containing, 104
TM-01-72-93 (NACE Standard Method), 176
Total organic carbon (TOC), 66–67
Total organic halide (TOX), 66–67
Transfer lines and hoses, 228
Types of SAE and AISI steel, 212–213

U Fuel, cetane number, 95
UN number, 234
Uniform corrosion, 152
Unwashed gum, 158, 171
Uses for steel, 211–212

Vacuum distillation, 6–7
Vacuum gas oils, 7, 48, 88, 136
Vanadium:
 in crude oil, 25–26, 32
 enhanced corrosion, 62
 in residual fuel, 62
Vane pump, 224
Vapor pressure, of gasoline components, 37
Vaporizing pot burner, 60
Varnish, 97, 103
Vehicle fuel tanks, 229
Vinyl acetate-fumarate copolymers, 145
Visbreaking, 9
Viscosity:
 effect on diesel fuel combustion, 52
 of residual fuel oils, 61
Viscosity limits for fuel oils, 108
Viton, 222
Volatility:
 of aviation gasoline, 42
 effect of low volatility fuel, 88–90
 effect on knock in gasoline engine, 130–131
 of gasoline, 37
Volumetric fuel consumption, effect of fuel density, 118
VV-F-800 (Federal specification), 51, 156

Waring blender test, 180

Washed gum, 171
Water:
 contamination, as a cause of rusting, 201
 electrical conductivity, 66
 emulsification of detergents, 69
 haze removal, 22–23, 140–142
 as a source of fuel problems, 67–69
 solvent properties, 65–67
Water and contaminants, sources, 139–140
Water and particulate removal, in fuel refining, 22–23
Water and sediment, in residual fuel, 63
Water in fuel systems, typical problems, 67–69
Water reaction:
 of aviation fuels, 178–179
 of aviation gasoline, 43
Water separability of petroleum oils and synthetic fluids, 179–180
Water separation index, modified, 163–164, 180–181, 206
Water soluble and fuel soluble biocides, 143–144
Water washing, 20, 148, 201–203, 208
Wax:
 classifications, 79–80
 in crude oil, 27
 crystal modifiers, 85–87, 145–148, 165–166
 crystal lattice, 85, 195
 deposition, 255
 description of petroleum wax, 79–80
 filtration problems, 84–85
 in fuel, 80–88, 117
 precipitation, 76
 problems, 123, 193–197
 in residual fuel, 62, 64, 88
 test methods measuring the effect of, 182–189
Wax crystal modifiers:
 description, 85, 145–148
 detection in fuel, 87
 minimized performance, 196–197
 problems in using, 165–166
 proper application, 86
 in residual fuel, 188
Wear metals and other contaminants from lubricant systems, 104
Welding, 153
Wet gas, 7
White smoke, 76
Wick vaporizing burner, 60
Williams Brothers pipeline stability test for distillate fuel, 173–174
Winter grade diesel fuel oil, 51–52
WSIM (*see* Water separation index, modified)
Yield stress, 146

Zytel, 222